I0035985

Solutions Manual

Real World Mathematics

Yink Loong Len

May Han Thong

SriBooks

An Imprint of the Simplicity Research Institute, Singapore.
www.simplicitysg.net

Solutions Manual: Real World Mathematics.
Published by SRI Books,
an imprint of the Simplicity Research Institute, Singapore.
www.sribooks.simplicitysg.net

*Sri**Books***

Copyright © 2016 by Yink Loong Len, May Han Thong, and the publisher.

All rights reserved. No part of this book may be reproduced in any
form, stored in any information retrieval system, or transmitted by
any means, except as permitted by provisions of the Copyright Act,
without written permission of the publisher.

For bulk orders, special discounts, or to obtain customised versions of this book,
please contact SRI Books at **enquiry@simplicitysg.net**.

A CIP record for this book is available from the
National Library Board, Singapore.

Edition SRI-2016-2.

ISBN: $978-981-09-8198-3$ (pbook)
ISBN: $978-981-09-8199-0$ (ebook)

Contents

Preface

This manual contains solutions to questions (not included here) from the book *Real World Mathematics* by W. K. Ng and R. Parwani. The material here is suitable for high-schools and colleges.

Topics covered: exponents, logarithms, polynomial equations, rational functions, simultaneous equations, matrices, coordinate and plane geometry, trigonometry, calculus, vectors and complex numbers.

Conventions:

When units of length are not specified in a question, we usually do not insert them into the solution.

About the Authors

Len Yink Loong and Thong May Han are mathematics enthusiasts based in Singapore and Malaysia.

Acknowledgements

May Han thanks her family and friends for their support during the production of this book.

Yink Loong thanks Rajesh Parwani and May Han for the opportunity to contribute to this project.

Jan 2016, Singapore.

Feedback: enquiry@simplicitysg.net

Chapter 1

Polynomials and Rational Functions

Recall: For the quadratic equation $ax^2 + bx + c = 0$, the values of x are given by:

$$x = \frac{-b \pm \sqrt{b^2 - 4ac}}{2a}. \tag{1.1}$$

---

1. Each book is to be sold at $30.

   (a) To break even, the money from the sale of x books needs to equal the cost, $C(x)$, required to produce them, that is $C(x) = 10x + 400 = 30x$, which gives $x = 20$.

   (b) The profit (sales minus the cost) required is $10,000. Hence, $30x - (10x + 400) = 10000$, that is $x = 520$.

   (c) 20% profit above the cost means that the profit is $1.2C(x) = 30x$. Therefore, $x = 26.7 \approx 27$.

2. Volume of fluid displaced = volume of solid submerged.

   (a) Volume of A is $V_A = a^2 b$ and the volume of B is $V_B = 4/3\pi r^3$. Total volume V of water displaced is $V = \frac{2}{3}V_A + \frac{3}{4}V_B = \frac{2}{3}a^2 b + \pi r^3$.

3. The pressure $P = \rho g h$.

   (a) Here, $\rho = 1000\,\text{kg/m}^3$, $g = 9\,\text{m/s}^2$ and $h = 10\,\text{m}$. Therefore, $P = 1000 \times 9.8 \times 10 \approx 100,000\,\text{N/m}^2$.

   (b) Pressure is additive: The total pressure on the object located 10 m below the sea level is the sum of the pressures from the sea and the atmosphere. Hence, in addition to the pressure of 100,000 N/m^2 from sea water as calculated above (the density of sea water is practically 1000 kg/m^3), there is the pressure from the atmosphere, which is by definition, 1 atmospheric pressure (atm). It turns out that 1 atm is about 100,000 Nm^{-2} (101,325 Nm^{-2}, to be precise), so the total pressure on the object is $P_{\text{total}} \approx 2$ atm.

4. Since $F \propto 1/R^2$, it means that $F_2/F_1 = (R_1/R_2)^2$. For $F_2 = \dfrac{F_1}{2}$, we have $\dfrac{1}{2} = \left(\dfrac{R_1}{R_2}\right)^2$ and so $R_2 = \sqrt{2}R_1$.

1

5. The total resistance R in a circuit with two resistors R_1 and R_2 is given by

$$\frac{1}{R} = \frac{1}{R_1} + \frac{1}{R_2}. \tag{1.2}$$

(a) Since R, R_1 and R_2 are all positive, $\frac{1}{R} \geq \frac{1}{R_1}$ and $\frac{1}{R} \geq \frac{1}{R_2}$. Therefore, $R \leq R_1$ and $R \leq R_2$.

(b) It is known that $R \geq 0$. From Eq. (1.2), we can see that if R_1 is fixed, then $\frac{1}{R} \geq \frac{1}{R_1}$. This means that $R \leq R_1$. Hence $0 \leq R \leq R_1$.

(c) Re-arranging the terms: $\frac{1}{R} = \frac{R_2 + R_1}{R_1 R_2}$. Hence, $R = \frac{R_1 R_2}{(R_1 + R_2)}$.

6. The frequency is given by $f = v/(4L)$.

(a) For $v = 345$ m/s and $L = 0.2$ m, $f = 345/0.8 = 431.25$ Hz.

(b) The human hearing range is roughly from 20 Hz to 20,000 Hz, so the sound made in part (a) will probably be audible.

7. Equilibrium constant $k = \dfrac{x}{(a - x)(b - x)}$, $x < a$, $x < b$.

(a) Given that $k = 0.9$, $a = 0.3$ and $b = 0.4$,

$$0.9 = \frac{x}{(0.3 - x)(0.4 - x)}$$
$$x^2 - 1.63x + 0.108 = 0.$$

Using formula (1.1), $x = 0.069$, or $x = 1.561$. However, since we require $x < a$ and $x < b$, x can only be equal to 0.069.

8. The Golden Ratio ϕ is defined by $\phi \equiv \dfrac{b}{a} = \dfrac{a}{b - a} > 0$.

(a) As $\dfrac{a}{b - a} = \dfrac{1}{b/a - 1} = \dfrac{1}{\phi - 1}$, so we obtain $\phi^2 - \phi - 1 = 0$ from the defining equation. From Eq. (1.1), the positive solution is $\phi = \dfrac{1 + \sqrt{5}}{2}$.

9. The maximum height is expressed as $H = \dfrac{v^2}{2g} + 0.9$, where $g = 9.8 \, \text{m/s}^2$.

(a) Since $H = 6$ m, we get $6 = \dfrac{v^2}{19.6} + 0.9$, that is $v = 10$ m/s.

(b) Here, $v = 5$ m/s. Therefore, $H = \dfrac{25}{19.6} + 0.9 = 2.18$ m.

10. The $A0$ paper has an area of $xy = 1$ m^2. Therefore, $y = 1/x$.

(a) Given that $\dfrac{x}{y} = \dfrac{y}{x/2}$, $x^2 = 2y^2$. Since $xy = 1$, it means that $x^2 = \dfrac{2}{x^2}$, that is $x = 2^{1/4}$.

(b) From $A0$ to $A1$, x is divided into 2 to form the longer edge of $A1$. Hence, the dimensions of $A1$ is $x/2$ by y. Subsequently, $A2$ is formed by dividing $A1$ along the y edge, giving $A2$ the dimensions of $x/2$ by $y/2$. Since the division alternates between the two edges, from $A2$ to $A4$ we would have divided each of the edges one more time to give $A4$ the dimensions $x/4$ and $y/4$.

From (a) we have found that $x = 2^{1/4}$ and that $y = 1/x = 2^{-1/4}$. Hence, the dimensions of $A4$ will be $x/4 = 297$ mm by $y/4 = 210$ mm.

(c) The size of BN is the geometric mean of the sizes of $A(N-1)$ and AN. Therefore, the length l (width w) of $B4$ will be the square root of the product of the lengths (widths) of $A3$ and $A4$ paper sizes, that is $l_B = \sqrt{l_{A3} \times l_{A4}}$. The dimensions of the $A3$ paper is given by the width $x/4 = 297$ mm and length $y/2 = 420$ mm, while from part (b), we know $A4$ has a width of 210 mm and length of 270 mm. Therefore, the $B4$ paper size will have a length of 353 mm and a width of 250 mm.

11. Height of the ball is expressed by the equation $y(t) = y_0 + ut - \dfrac{gt^2}{2}$.

(a) Given that $y_0 = 5$ m, $u = 10$ m/s and $g = 10$ m/s^2, we obtain $0 = 5 + 10t - 5t^2$. Using the formula in (1.1), $t = 1 \pm \sqrt{2}$. Since $\sqrt{2} > 1$ and time cannot be negative in value in this case, the answer is $t = 1 + \sqrt{2}$.

(b) The velocity of the ball is zero at the point it reaches the maximum height h_{max}. The velocity can be found by differentiating $y(t)$ with respect to t, to obtain $v(t) = u - gt$. Hence, at maximum height, $v(t) = 10 - 10t = 0$, giving $t = 1$ s. Substituting $t = 1$ s into the expression for the height, we obtain $h_{max} = 10$ m.

(c) The expression for the instantaneous height of the ball is given by $y(t) = 0 = 5 + 10t - 5t^2$. Therefore $y(t) = 2$ m, when $5t^2 - 10t - 3 = 0$. Once again using (1.1), we end up with $t = 1 + \dfrac{2}{5}\sqrt{10}$.

(d) The value of $g_{moon} = (10/6)$ m/s^2. Hence, $0 = 5 + 10t - 5t^2/6$. Solving this gives us $t = 6 + 10\sqrt{42}$ s.

From (b), we know that $v(t) = u - gt = 0$ at the maximum height. Given that $u = 10$ m/s, the maximum height on the moon is reached at $t = 6$ s, with $y = 35$ m.

For $y = 2$ m, we obtain $5t^2 - 10t - 3 = 0$, which implies $t = 6(1 + \sqrt{110}/10)$ s.

12. $N(t) = N_0 - ct + at^2$, with $N_0 = 100, c = 5$, and $a = 1$.

(a) $N(t) = 200$. Hence, $t^2 - 5t - 100 = 0$. From (1.1), $t = 12.8$ s.

(b) Here, $N(t) = 94$, so $t^2 - 5t + 6 = 0$. Factoring the equation , we obtain $(t - 2)(t - 3) = 0$. Hence, $t = 2$ s or $t = 3$ s.

(c) For $N(t) = 93$, we get $t^2 - 5t + 7 = 0$, that is $t = (5 \pm \sqrt{-3})/2$. The solution contains an imaginary value, and since time should take a positive value, this solution is invalid. This means that $N(t)$ cannot take the value of 93.

(d) With the addition of the bt^3 term, where $b = 2$, the formula becomes $N(t) = 100 - 5t + t^2 + 2t^3$.

   i. At the point where $N(t)$ is minimum, its derivative must vanish, $dN/dt = 0$, and second derivative must be positive, $d^2N/dt^2 > 0$. We have

$$\frac{dN}{dt} = 6t^2 + 2t - 5 = 0, \tag{1.3}$$

$$\frac{d^2N}{dt^2} = 12t + 2 > 0, \tag{1.4}$$

   since $t \geq 0$. Hence, we have confirmed that this turning point is indeed the minimum. To find the value of t where this minimum occurs, we solve Eq.(1.3) to obtain $t = 0.76$ s. Therefore, $N(0.76) = 98$. (A sketch confirms this to be a global minimum for $t \geq 0$).

13. The trajectory is expressed as $y = ax - bx^2$, with the mouth of the cannon being at the origin, $(x = 0, y = 0)$, and the ground at $y = -1.5$ m. The ball reaches a maximum height of $51/4$ m, and hits the ground at $x = 80.46$ m.

(a) Let x' be the value of x when the ball reaches maximum height. From the information given, we obtain two equations:

$$-1.5 = a(80.46) - b(80.46)^2, \tag{1.5}$$

$$\frac{51}{4} - 1.5 = ax' - bx'^2. \tag{1.6}$$

At maximum height, $dy/dx = a - 2bx' = 0$. Therefore, $x' = \dfrac{a}{2b}$. Substituting this into Eq. (1.6), we obtain

$$\frac{51}{4} - 1.5 = \frac{45}{4} = \frac{a^2}{2b} - \frac{a^2}{4b} \Rightarrow 45 = \frac{a^2}{b}. \tag{1.7}$$

Putting the result from Eq. (1.7) into (1.5), we obtain a quadratic equation:

$$\frac{(80.46)^2}{45}a^2 - 80.46a - 1.5 = 0,$$

which can be solved using the quadratic formula (1.1). From this, we obtain $a = 0.577$ and $b = 0.0074$.

(b) For the maximum point, $dy/dx = a - 2bx = 0$, so $a = 2bx$. From part (a) we know that $a = 0.58$ and $b = 0.0074$, so $x = 39$. Consequently, $y = 11$ m.

(c) Setting $y = 0.58x - 0.0074x^2$ to zero gives $x = 0$ and $x = 78$ m.

(d) For $y = 10$, we obtain $-0.0074x^2 + 0.58x - 10 = 0$, which implies $x = 26$ m or $x = 52$ m. This means that for $26 \le x \le 52$, y will be above 10 m.

14. When $p = T = 1$, the Van der Waals equation is

$$\left(1 + \frac{3}{v^2}\right)\left(v - \frac{1}{3}\right) = \frac{8}{3}.$$

(a) Re-arranging the terms, we get $(v^2 + 3)(3v - 1) - 8v^2 = 0$. Expanding and simplifying the last equation gives $(v - 1)^3 = 0$, implying $v = 1$.

(b) Re-arranging the terms: $p = -\dfrac{3}{v^2} + \dfrac{8T}{3}\left(\dfrac{1}{v - 1/3}\right)$.

15. Volume of the fuel is given by $V = \dfrac{\pi h^2}{3}(3R - h)$.

(a) When $h = 2R$, the tank has been filled to the top. The volume of the fuel is then $V = \dfrac{4\pi R^3}{3}$, which is indeed the volume of a sphere.

(b) If $R = 5$ and $V = 36\pi$, then $f(h) \equiv h^3 - 15h^2 + 108 = 0$. We easily see that $f(0)$ is positive while $f(5)$ is negative. So there must be at least one root between $h = 0$ and $h = 5$. Trying integral values, we find that $h = 3$ is a root.

(c) From part (b), we know that one of the roots for the equation is $h = 3$. Therefore, $f(h) = (h - 3)(ah^2 + bh + c) = 0$, leaving us the task of finding the values of a, b, and c. Using long division, we find that $a = 1, b = -12$, and $c = -36$, that is, $h^2 - 12h - 36 = 0$ determines the other roots. From this, through the application of the quadratic formula, we find $h = 6(1 \pm \sqrt{2})$, of which only the solution with the '+' sign is tentatively acceptable.

However, another condition that needs to be taken into consideration is the fact that $h < 2R$. Since $R = 5$, this means that from the two solutions $h = 3$ and $h = 6(1 + \sqrt{2})$, only $h = 3$ is physically permissible.

(d) Volume, by definition, is the space enclosed by a certain closed surface. If we single out a particular direction as height, volume can also be thought as sum of layers of cross-sectional area, each of infinitesimal height. When the cross-sectional areas are independent of the height, we have the usual "area × height" formula; popular examples are volumes of cubes and cylinders. In general, the cross-sectional area could be dependent on the height, as is the situation in the exercise.

Fig. (1.1) show the side-view of the spherical tank of radius R, with fuel filled up to height h. It is clear that the cross-section of a sphere is a circle (imagine viewing the fuel from the top of the spherical tank). Hence, for any height h', the cross-sectional area is $A(h') = \pi r^2$, with r^2 given by $R^2 - (R - h')^2 = 2Rh' - h'^2$.

The volume of fuel filled up to h is then the "sum", or integral,

$$V(h) = \int_0^h A(h')\,dh' = \int_0^h \pi(2Rh' - h'^2)\,dh' = \frac{\pi h^2}{3}(3R - h).$$

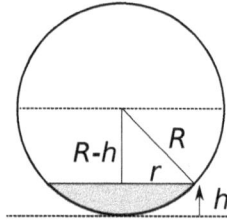

Figure 1.1: Side-view of the spherical tank.

16. The box has a square base with sides of length $(a - 2x)$, and height x. Therefore, the volume of the box is $V(x) = x(a - 2x)^2$. (Note: Both x and a are positive.)

   (a) As physically $V \geq 0$, and $(a - 2x)^2 \geq 0$, so we see by inspection that the minimum volume corresponds to the situation $x = 0$ (no height) or $x = a/2$ (no base area). Both minima correspond to zero volume.

   (b) As $V(x) \geq 0$ is a smooth function, it must have a maximum in the physical range $0 < x < a/2$, since it vanishes at the ends. The turning points satisfy the equation

   $$\frac{dV}{dx} = a^2 - 8ax + 12x^2 = 0,$$

   which gives us the values $x = a/2$ and $x = a/6$. To determine the nature of the extrema (minima or maxima), we need to check the value of $d^2V/dx^2 = -8a + 24x$. For $x = a/2$, we have $d^2V/dx^2 > 0$, so this is a minimum point (as we already knew from part (a)). For the other extremum at $x = a/6$, we have $d^2V/dx^2 = -4a$, so $x = a/6$ is at least a local maximum. Since at the boundaries $x = 0$ and $x = a/2$, the smooth function $V(x)$ does not exceed its value at the local maximum, so the local maximum is actually a global maximum.

(c) The plot of V against x is shown as in Fig. (1.2).

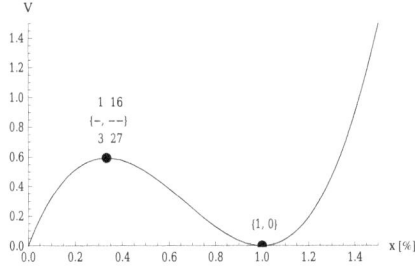

Figure 1.2: Graph of V as a function of x. The maximum point is at $x = 1/3$, with $V = 16/27$. (Note: The physical region is $0 \le x \le 1$).

(d) The volume is given by $V = x(a - 2x)^2$. From (b) we know that the maximum volume is obtained when $x = a/6$. When $a = 2$, $x = 1/3$ and $V = 16/27$.

17. Relative velocity $v' = \dfrac{v - u}{1 - uv/c^2}$.

(a) Rearranging:

$$v - u = v'\left(1 - \frac{uv}{c^2}\right)$$

$$v' + u = v\left(1 + \frac{uv'}{c^2}\right) \Rightarrow v = \frac{v' + u}{\left(1 + \frac{uv'}{c^2}\right)}.$$

(b) When $v = c$, $v' = \dfrac{c - u}{1 - u/c} = c$.

(c) For $v = v'$, we get the expression $vc^2 - uv^2 = vc^2 - uc^2$, that is $v^2 = c^2$. Hence, $v = \pm c$.

(d) When $v/c \ll 1$ and $u/c \ll 1$, the value of $1 - uv'/c^2 \approx 1$. Therefore, $v' \approx v - u$.

18. Given that $\dfrac{1}{u} + \dfrac{1}{v} = \dfrac{1}{f}$,

(a) Rearranging:

$$\frac{1}{v} = \frac{1}{f} - \frac{1}{u} = \frac{u - f}{uf}$$

$$\Rightarrow v = \frac{uf}{u - f} = \frac{(u - f)f + f^2}{u - f} = f + \frac{f^2}{u - f}.$$

(b) The second term in the above expression is singular at $u = f$. For $u = f^+$ (that is, u slightly larger than f) we have $v \to +\infty$ while for $u = f^-$ we have $v \to -\infty$. This means that the range of v is given by $-\infty \leq v \leq \infty$.

19. The function is expressed as $f(x) = 2x^2 - 3x + 1$.

   (a) At the turning point, $df(x)/dx = 4x - 3 = 0$. Hence, $x = 3/4$ and $f(3/4) = 1/8$. Differentiating $f(x)$ for the second time, we obtain $d^2 f(x)/dx = 4$. This means that the turning point is a local minimum.

20. The amount of cement at a time t is given by $V(t) = 5t^2 - \dfrac{t^3}{3} + 10$.

   (a) The graph of $V(t)$ against t is as shown in Fig. (1.3). It can be seen from Fig. (1.4) that $V(t)$ goes to zero in the region $15 < T < 15.5$.

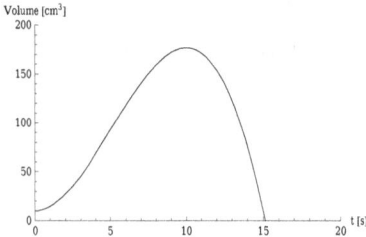

Figure 1.3: Graph of $V(t)$ against t.

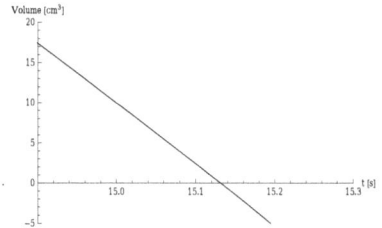

Figure 1.4: $V(t)$ near its zero.

   (b) Since $T = 15 + x$, with $0 < x < 0.5$, and $V(t) = 0$,

$$5(15 + x)^2 - \frac{(15 + x)^3}{3} + 10 = 0$$

$$-\frac{x^3}{3} - 10x^2 - 75x + 10 = 0.$$

   Ignoring terms that are quadratic or higher order in x, we obtain $10 - 75x \approx 0$ for x small, that is for $0 \leq x \leq 0.5$.

   (c) From part (b), we found that $10 - 75x \approx 0$. Hence, $x \approx 0.1333$. As $T = 15 + x$, we now know that $T \approx 15.1333$.

   (d) If in part (a) we keep terms quadratic in x, we will obtain $2x^2 + 15x - 2 \approx 0$, which implies $x \approx 0.13104$ and $T \approx 15.13104$.

   (e) Using computer software, the value of T is determined to be 15.13103.

Chapter 2

Exponents and Logarithms

1. In standard notation, a quantity is expressed in the form $A \times 10^n$ with $1 \leq |A| < 10$; see Section 11.1 of the main text. Therefore, $c = 299792458 \, \text{m/s}$ can be expressed as $c = 2.99792458 \times 10^8 \, \text{m/s}$.

2. Denoting the distance between the Earth and the Moon as x, we have $2x = c \times 2.6 \, \text{s}$. Up to two significant figures, we then have $x = 3 \times 10^8 \times 1.3 = 3.9 \times 10^8 \, \text{m}$.

3. (a) Light travels about 3×10^8 m in a second, while in a year of 365 days we have $365 \times 24 \times 3600$ seconds. A light-year is thus $3 \times 10^8 \times 365 \times 24 \times 3600 = 9.46 \times 10^{12}$ km.

   (b) It takes 6 years for light to reach the planet. For the spaceship which moves at the $1/5$ times the speed of light, the time required is then 5 times longer, that is, $6 \times 5 = 30$ years.

   (c) 18 years is 3 times of 6 years. Hence, the speed of the spaceship is 3 times slower, that is $c/3$.

   (d) The distance travelled in 12 years with speed $c/3$ is 4 light-years. In total, the spaceship has travelled 10 light-years.

4. (a) 18 grammes of water contain 6.02×10^{23} molecules, so 1 gram contains $6.02 \times 10^{23}/18$ molecules. With the density of water as $1 \, \text{kg/litre}$, 1 litre contains 1 kg of water. Thus, there are $1000 \times 6.02 \times 10^{23}/18 = 3.34 \times 10^{25}$ water molecules in 1 litre.

   (b) 1 gram contains $6.02 \times 10^{23}/18$ molecules, so 1 molecule weighs $\dfrac{18}{6.02 \times 10^{23}}$ g. 1000 water molecules therefore weigh $1000 \times 18/(6.02 \times 10^{23}) = 3 \times 10^{-20}$ g.

5. (a) A straight line has the form $y = mx + c$, where m is the slope and c is the y-axis intercept. What the student found is $\log T = \dfrac{1}{2} \log l + c$. Writing c as $\log d$ for some d (which is possible for all finite and real c), and using the identities $\log a + \log d = \log(ad)$ and $a \log d = \log(d^a)$, we have $\log T = \log(\sqrt{l}d)$. Comparing both sides, we get $T = d\sqrt{l}$, that is $T \propto \sqrt{l}$.

   (b) If $l \to 2l$, $T = d\sqrt{2l}$, and so $T \to \sqrt{2}T$.

6. (a) When $v = 0$, $1 - v^2/c^2 = 1$, hence $E = mc^2$.

   (b) For $m \neq 0$, E is finite and positive as long as $1 - v^2/c^2$ is not zero. This implies that the speed limit for an object with non-zero mass is c.

(c) Re-arranging the formula, we get $E = mc^2(1 - v^2/c^2)^{-1/2}$. Next, recall (see Appendix A of the main text) the Binomial expansion $(1 + x)^n = 1 + nx + \dfrac{n(n-1)}{2!}x + \dfrac{n(n-1)(n-2)}{3!}x^2 + \cdots$, where for small x, expansion up to the second term is usually sufficient. Then, for $v/c \ll 1$, we have $E \approx mc^2(1 + \dfrac{1}{2}v^2/c^2) = mc^2 + \dfrac{1}{2}mv^2$, which is the well-known expression for rest mass energy plus the non-relativistic kinetic energy.

7. (a) $v = \sqrt{9.8 \times 2000} = 140 \,\text{m/s}$.

   (b) $v/3 = \sqrt{g\dfrac{d}{9}}$, that is the depth decreases by a factor of 9.

8. (a) After 40 hours, the yeast multiplied so that $N(40) = 2N(0)$, where $N(0)$ is its original quantity at $t = 0$. Similarly, we have $N(80) = 2N(40) = 4N(0)$, and $N(120) = 2N(80) = 8N(0)$.

   (b) From the pattern above, we see that $N(k \times 40) = 2^k N(0)$. For a general time t, we thus have $N(t) = 2^{t/40}N(0)$.

   (c) For $N(0) = 5\,\text{g}$, we have $N(20) = \sqrt{2}N(0) = 7.07\,\text{g}$.

   (d) For $N(t) = 30 = 2^{t/40} \times 10$, we have $2^{t/40} = 3$, so that $t = 40\log(3)/\log(2) = 63.4$ hours.

9. (a) This is similar to Exercise 2.8 above, with 2 years replacing 40 hours for the characteristic time. Thus, $N(t) = 2^{t/2}N(0)$, where t is measured in years.

   (b) For $N(t) = 20N(0) = 2^{t/2}N(0)$, we have $\ln(20) = \ln(2^{t/2}) = \dfrac{t}{2}\ln(2)$, such that $t = 2\ln(20)/\ln(2) = 8.64$ years.

10. (a) The plot is shown in Fig.(2.1).

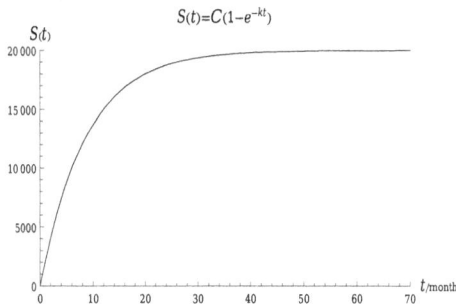

Figure 2.1: Plot of $S(t)$, with $C = 20,000$, and $k = \ln(2)/6$.

(b) From the graph above, it is obvious that the contribution from the current monthly sales is negligible when kt becomes large, since the time-dependent term e^{-kt} is then very small and $S(t)$ becomes almost constant, approaching the value of C. Hence, $C \approx 20000$ is the total sales volume after a long time period.

(c) Half of the total sales is $C/2$. Solving $\dfrac{C}{2} = C(1 - e^{-6k})$ then gives $e^{-6k} = 1/2$, that is $k = \dfrac{\ln 2}{6}$.

(d) 1 year is 12 months, so $S(t) = 20,000(1 - e^{-2\ln 2}) = 20,000 \times 3/4 = 15,000$, since $e^{-2\ln 2} = e^{-\ln 4} = 1/4$.

11. (a) $\dfrac{N_0}{2} = N_0 e^{-\lambda t}$ implies $\dfrac{1}{2} = e^{-\lambda t_{1/2}}$, that is $\lambda t_{1/2} = \dfrac{\ln 2}{\lambda}$.

(b) $\dfrac{1}{5} = e^{-\frac{\ln 2}{5730}t}$, that is $t = \dfrac{5730 \ln 5}{\ln 2} = 13305$ years.

(c) $0.12 = e^{-\lambda \times 7500}$, so $\lambda = -\dfrac{\ln 0.12}{75000}$. Then, $t_{1/2} = -\dfrac{75000 \ln 2}{\ln 0.12} = 24519 \approx$ 24500 years.

(d) Instead of writing $N = N_0 e^{-\lambda t}$, we wish to write $N = N_0 (\dfrac{1}{2})^{-\gamma t}$ for some γ. Since $N = N_0/2$ for $t = t_{1/2}$, we immediately have $\gamma = -1/t_{1/2}$.

More systematically, we want $e^{-\lambda t} = (\dfrac{1}{2})^{-\gamma t}$, so, by taking the log of both sides, we get $-\lambda t = -\gamma t \ln(1/2)$, or $-\lambda t = \gamma t \ln 2$. Since $\lambda = \ln 2/t_{1/2}$, we recover $\gamma = -1/t_{1/2}$.

12. (a) Projected revenue in 3 years is $100\,\mathrm{M} \times 1.05 \times 1.05 \times 1.05 = 100\mathrm{M} \times (1.05)^3 = 115.76\,\mathrm{M}$.

(b) $200\,\mathrm{M} = 100\,\mathrm{M} \times (1.05)^x$. Equivalently, $\ln(2) = x \ln(1.05)$, that is $x = 14.2$, in years.

13. (a) Neutral solution means pH=7, so $7 = -\log_{10}[H^+]$, or $[H^+] = 10^{-7}$ moles per litre.

(b) We have $[H^+] = 10^{-(\mathrm{pH})}$. Since acidic solutions have a pH less than 7, so they will have a higher hydrogen ion concentration than a neutral solution.

(c) Since its hydrogen ion concentration is larger than pure water, it is acidic. The pH value is $-\log_{10}(5 \times 10^{-7}) = -\log_{10} 5 + 7 = 6.3$.

14. (a) $30 = 10 \log_{10}(I/I_0)$, that is $I/I_0 = 1000$.

(b) $L = 10 \log_{10}(4) = 6$.

15. (a) $8.3 = \log_{10}(I/I_0)$, that is $I/I_0 = 10^{8.3} = 1.995 \times 10^8$.

(b) $R = \log_{10}(10^7) = 7$.

(c) First write $\log_{10} \dfrac{U}{U_0} = \dfrac{3R(U)}{2}$. Then, $\log_{10} \dfrac{U_1}{U_2} = \log_{10} \dfrac{U_1}{U_0} - \log_{10} \dfrac{U_2}{U_0} = \dfrac{3(R(U_1) - R(U_2))}{2}$, which for the two earthquakes above is equal to 1.95.

Thus $\dfrac{U_1}{U_2} = 10^{1.95} = 89$.

16. (a) For the Lyman series ($n_1 = 1$), we have $\dfrac{1}{\lambda} = R\left(1 - \dfrac{1}{n_2^2}\right)$. For shortest wavelength, we require the right-hand side to be as large as possible, which is satisfied for $n_2 \to \infty$. Thus $\lambda = 1/R = 9.12 \times 10^{-8}$ m.

(b) For the Balmer series we have $\dfrac{1}{\lambda} = R\left(\dfrac{1}{2^2} - \dfrac{1}{n_2^2}\right)$. For longest wavelength, we require the right-hand side to be as small as possible, which is satisfied for $n_2 = 3$. Thus $\lambda = 36/(5R) = 6.56 \times 10^{-7}$ m.

(c) $\dfrac{1}{\lambda} = R\left(\dfrac{1}{n_1^2} - \dfrac{1}{n_2^2}\right) = R\left(\dfrac{n_2^2 - n_1^2}{n_1^2 n_2^2}\right)$, and so $\lambda = \dfrac{n_1^2 n_2^2}{R(n_2^2 - n_1^2)}$.

17. (a) For convenience, we first define the hyperbolic cosine function $\cosh(x) \equiv \left(\dfrac{e^x + e^{-x}}{2}\right)$. Then, we have $y = a\cosh(x/a)$. Similarly, we define $z \equiv a\sinh(x/a)$, where $\sinh(x) \equiv \left(\dfrac{e^x - e^{-x}}{2}\right)$ is the hyperbolic sine function. It can be easily verified that $\dfrac{d\cosh(x)}{dx} = \sinh(x)$. The minimum point of y is at where $\dfrac{dy}{dx} = 0$, which is satisfied only when $x = 0$. Therefore, we have $y = a = 10$ m at the minimum. The support level is 100 m above the minimum, that is, at $y = 110$ m.

The x-coordinates of the two support points are obtained by solving $110 = 10\cosh(x/10)$. Explicitly, $x = \pm 10\cosh^{-1}(11) = \pm 30.89$, where $\cosh^{-1}()$ is the inverse hyperbolic cosine function. The $\pm$ sign comes from the fact that y is unchanged upon the substitution $x \to -x$.

(b) The bottom of the catenary is at $x = 0$. Near its bottom, we can thus expand y as $y = \dfrac{(1 + x + x^2/2 \cdots) + (1 - x + x^2/2 + \cdots)}{2a} \approx \dfrac{2 + x^2}{2a}$, which is a parabola.

18. The total mass is given by "volume × density", that is $Am = \dfrac{4}{3}\pi R^3 \rho$. Hence, $R = \left(\dfrac{3m}{4\pi\rho}\right)^{1/3} A^{1/3}$.

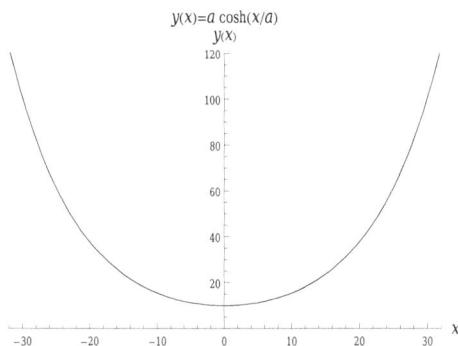

$$y(x)=a \cosh(x/a)$$

Figure 2.2: Plot of $y(x)$, with $a = 10$.

19. (a) Using the Binomial expansion, we have

$$\$10000 \times (1 + 0.02)^6$$
$$= \$10000 \times [1 + 6(0.02) + 15(0.02)^2 + 20(0.02)^3 + 15(0.02)^4$$
$$+ 6(0.02)^5 + (0.02)^6]$$
$$= \$11262.$$

This is of course confirmed by a direct calculation using a calculator.

(b) $(1 + 0.02)^x = 1.2$, that is $x = \ln 1.2 / \ln 1.02 = 9.2$, which is rounded to 10 months.

20. $\left(1 + \dfrac{0.05}{12}\right)^{12} = 1.0512$, which means the equivalent annual simple interest is 5.12%.

21. $15000 = x\left(1 + \dfrac{0.03}{4}\right)^8$, that is $x = 15000/1.0616 = 14130$.

22. $20000 = 12000\left(1 + \dfrac{R}{4 \times 100}\right)^{12}$, that is $\ln 5 - \ln 3 = 12 \ln\left(1 + \dfrac{R}{400}\right)$, such that $1 + \dfrac{R}{400} = e^{(\ln 5 - \ln 3)/12}$, and $R = 400\, e^{(\ln 5 - \ln 3)/12} - 400 = 17.4$.

23. (a) Since T_a is a constant, for $y \equiv T - T_a$, we have $\dfrac{dy}{dt} = \dfrac{dT}{dt} = -k(T - T_a) = -ky$.

(b) Using $\dfrac{d}{dt} e^{kt} = k e^{kt}$, one can immediately verify that $y = A e^{-kt}$ satisfies $\dfrac{dy}{dt} = -ky$.

(c) $T(0) = T_i$ implies $-kA = -k(T_i - T_a)$, since $e^0 = 1$. Thus, $A = T_i - T_a$, and $T(t) = y + T_a = (T_i - T_a)e^{-kt} + T_a$.

(d) When $t = 0$, we get $T(0) = T_i$, indeed. When $t \to \infty$, $e^{-kt} \to 0$, so $T(\infty) \to T_a$, which is also as expected: This means the object reaches thermal equilibrium with its environment.

(e) $(100 - 25)e^{-0.06t} = 50 - 25$, that is $e^{-0.06t} = 1/3$, $t = \ln 3/0.06 = 18.31$, in minutes.

(f) $x = 25 + (100 - 25)e^{-0.06(15+\ln 3/0.06)} = 35.2$, in degree Celsius.

24. (a) $v(t) = \dfrac{g}{b}(1 - e^{-bt}) \Rightarrow \dfrac{dv(t)}{dt} = -\dfrac{g}{b}(-be^{-bt}) = ge^{-bt} = g - bv(t)$. Also, $v(0) = g \times (1 - 1) = 0$.

(b) As t gets very large, $e^{-bt} \to 0$ and we have $\displaystyle\lim_{t\to\infty} v(t) = \dfrac{g}{b}$.

(c) $\dfrac{dv}{dt} = 0 = g - bv$ implies $v = g/b$, as above.

(d) First we re-write the differential equation as $\left(\dfrac{d}{dt} + b\right)v(t) = g$. Then, note that the differentiation $\left(\dfrac{d}{dt} + b\right)v(t)$ is identical to $e^{-bt}\dfrac{d}{dt}\left(e^{bt}v(t)\right)$, as can be verified using integration by parts. Multiplying e^{bt} on both sides, we have $\dfrac{d}{dt}\left(e^{bt}v(t)\right) = e^{bt}g$, and a direct integration gives us $e^{bt}v(t) = \displaystyle\int_0^t e^{bt'} g\, dt' + e^{bt}v(t)\Big|_{t=0}$. For $v(0) = 0$, we then have $e^{bt}v(t) = \dfrac{g}{b}(e^{bt} - 1)$, and finally $v(t) = \dfrac{g}{b}(1 - e^{-bt})$.

(e) It is important for parachutists to reach a low terminal velocity so that they hit the ground safely.

25. (a) $\rho(r) = \dfrac{\rho_0}{1 + e^{\frac{r-R}{a}}}$. So, when $r = R$, $\rho(R) = \dfrac{\rho_0}{1 + e^0} = \dfrac{\rho_0}{2}$.

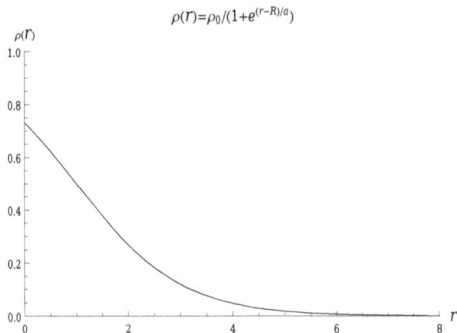

Figure 2.3: Plot of $\rho(r)$, with $\rho_0 = 1$, $R = 1$, and $a = 1$.

(b) We are looking for $x \equiv r_1 - r_2$, where $\rho(r_1) = \rho_0/10$, and $\rho(r_2) = 9\rho_0/10$. This means $e^{\frac{r_1-R}{a}} = 9$ and $e^{\frac{r_2-R}{a}} = 1/9$, which is satisfied by $r_1 = R + a\ln 9$ and $r_2 = R - a\ln 9$. Thus, $x = 2a\ln 9$, and finally $a = \dfrac{x}{2\ln 9}$.

(c) If $x = 2.4\,\text{fm}$, then $a = \dfrac{2.4}{2\ln 9} = 0.546\,\text{fm}$.

**Other titles by SRI Books and
the Simplicity Research Institute**

Integrated Mathematics for Explorers by Adeline Ng and R. Parwani
A book for mathematics lovers in school, college and beyond. Includes exercises, problems, challenges, investigations, puzzles, unsolved mathematical problems, and beautiful theorems.

Real World Mathematics by Wei Khim Ng and Rajesh R. Parwani
A resource for educators and learners interested in real-world applications of mathematics. Suitable for high schools and colleges.

*Simplicity in Complexity: An Introduction to Complex Systems
by Rajesh R. Parwani*
Introduces self-organisation, emergence, agent-based simulations, complex networks, phase plane plots, fractals, chaos, and measures of complexity. Emphasis is placed on clarifying common misconceptions.

Available from **www.store.simplicitysg.net** and other outlets.

Previews at **www.simplicitysg.net/books**.

Chapter 3

Simultaneous Equations and Matrices

1. Temperature readings in the two scales are related by $F = aC + b$.

   (a) When $C = 0$, the value of F is 32. Hence, $b = 32$. Then, given that $F = 212$ when $C = 100$, we deduce $a = 9/5$.

   (b) Re-arranging gives $C = (F - b)/a$. Hence, $C = 5(F - 32)/9$.

   (c) For $C = F$, we have $C = \frac{9}{5}C + 32$. Therefore, $C = -40$.

   (d) Substituting $C = 5(F - 32)/9$ into $T = C + 273.15$, we obtain $F = 1.8T - 459.67$.

   (e) For $T = 0$, $C = -273.15$ and $F = -459.67$.

2. The maximum amount of weight that you can carry has been stated to be 8 kg per trip. Since a bag of grapefruits weighs 5 kg, this means that no matter the combination, you can only carry 1 bag of grapefruits per trip. Subsequently, with respect to the weight of one bag of apples (2kg) and the maximum load you can take (8 kg), the only combination that can work with a bag of grapefruits will be one bag of grapefruits and one bag of apples. As there are 6 bags of grapefruits, this means that you will need 6 trips to transport all the grapefruits, together with 6 bags of apples. Accomplishing this will leave you with the remaining 6 bags of apples.

   Now, each trip will allow you to carry up to 4 bags of apples. The remaining 6 bags of apples will then require an additional two trips. Therefore, the total number of trips required is eight.

3. Since the car has to travel 300 km in less than 6 hours, the lower limit of v is given by $v > 300/6 = 50$ km/h. The fuel consumption rate is $C = (v/20)^3$. The total fuel consumption after t hours is $F = C \times t = Cd/v$ where $d = 300$ km. We require $F < 150$, which implies $v < 63.25$.

4. Equating $F = mg$ and $F = GMm/R^2$ gives $g = GM/R^2$. Putting in the given values of g, R and G, we obtain $M = 5.96 \times 10^{24}$ kg.

5. The supply is expressed as $Q_s = 2P - 10$, while the demand is $Q_d = 25 - 2P$.

   (a) Inverting the equations, we obtain $P = Q_s/2 + 5$ and $P = (25 - Q_d)/2$.

   (b) Please refer to Figure 3.1.

   (c) For equilibrium we set $Q_s = Q_d$ to get $P = 35/4$ and $Q = 15/2$.

   (d) When the price goes above the equilibrium condition, that is $P > 35/4$, Q_d decreases while Q_s increases. Therefore, there will be an excess supply of goods, exceeding the demand.

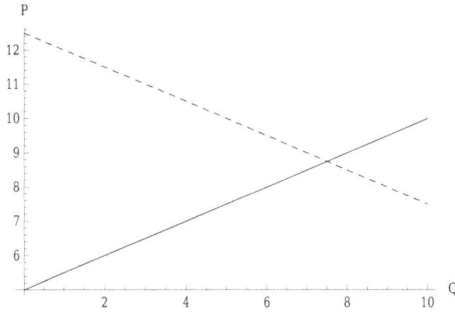

Figure 3.1: Graph of P vs. Q for the two expressions.

6. Given that $t' = t$, $x' = x - ut = x - ut'$. Inverting the last equation, we get $x = x' + ut'$.

7. The Lorentz transformations are:

$$t' = \gamma(t - ux/c^2) \quad \text{and} \quad x' = \gamma(x - ut) \ .$$

   (a) In matrix form, the Lorentz transformation can be expressed as

$$\begin{pmatrix} t' \\ x' \end{pmatrix} = \begin{pmatrix} \gamma & -\gamma u/c^2 \\ -\gamma u & \gamma \end{pmatrix} \begin{pmatrix} t \\ x \end{pmatrix} \equiv I \begin{pmatrix} t \\ x \end{pmatrix}$$

   (b) The inverse matrix I^{-1} is (note $\gamma^2(1 - u^2/c^2) = 1$),

$$I^{-1} = \begin{pmatrix} \gamma & \gamma u/c^2 \\ \gamma u & \gamma \end{pmatrix}$$

   Hence,

$$\begin{pmatrix} t \\ x \end{pmatrix} = I^{-1} \begin{pmatrix} t' \\ x' \end{pmatrix}$$

   giving $t = \gamma(t' + ux'/c^2)$ and $x = \gamma(x' + ut')$.

   (c) For $u \ll c$, $u/c^2 \approx 0$ and $\gamma = 1$. Therefore, $t' = t$ and $x' = x - ut$.

8. At terminal velocity V_T, the force due to the gravitational pull, $F = mg$, is balanced by the force due to the air-resistance, $F = kA(V_T)^2$. Equating the two expressions for F we get $V_T = \sqrt{mg/kA}$.

9. (a) To minimize travel time, we can attempt to maximize the travelling speed. This means that we have to maximize the running time, that is 3 hours. However, as each hour of running must be accompanied by 20 minutes of rest, the last part of the journey should be covered through running such that the resting time could be spent after reaching town B. 3 hours of running would cover 18 km, so the remaining 24 km needs to be covered by walking ($= 24/3 = 8$ hr.)

(b) Total time, t_{total} is the sum of the time spent on running (3 hours), with 2 rest breaks (40 minutes) and the time spent on walking (8 hours). Therefore, $t_{total} = 11$ hours 40 minutes. (We leave you to check that trying to avoid the 20 min rest breaks by doing more walking rather than running does not shorten the overall time taken).

10. The upward (buoyancy) force is $F_u = \rho_0 V_0 g$.

(a) When $W = F_u$, we obtain $\rho_0 V_0 g = \rho V g$. Therefore $\rho = \rho_0 V_0 / V$.

(b) The volumes of the blocks A and B are $a^2 b$ and $4\pi r^3/3$, respectively. Also, we have found $V_0 = 2a^2 b/3$ for block A and $V_0 = \pi r^3$ for block B. Since $\rho = \rho_0 V_0 / V$, we get $\rho_A = 2\rho_0/3$ and $\rho_B = 3\rho_0/4$.

11. The given information can be summarised by the two simultaneous equations

$$x + y + z = 1,000,000 , \tag{3.1}$$
$$0.06x + 0.08y + 0.10z = 70,000 . \tag{3.2}$$

(a) We first eliminate y from the above equations by subtracting (3.2) from $0.08\times$ Eq.(3.1) to get

$$z = x - 500,000 . \tag{3.3}$$

Substituting (3.3) into (3.2), we obtain

$$y = 1,500,000 - 2x . \tag{3.4}$$

Hence, the amount borrowed is x at 6%, $y = 1,500,000 - 2x$ at 8% and $z = x - 500,000$ at 10%.

(b) Each of x, y and z should be equal or more than 0. Hence, from $y \geq 0$, we obtain $x \leq 750,000$, while $z \geq 0$ gives us $x \geq 500,000$. This results in $500,000 \leq x \leq 750,000$.

(c) The director borrowed \$ 200,000 more at 8% than at 10%, which means that $y = z + 200,000$. From part (a), we also know that $z = x - 500,000$, so $y = x - 300,000$. Substituting this into (3.4) gives $x = 600,000$. Therefore, $z = x - 500,000 = 100,000$, and $y = 1,500,000 - 2x = 300,000$.

12. The total number of teams, N, is the sum of x and y, which represent the number of teams with 3 and 5 members respectively. The total number of members is 210. The scenario is summarised by the following equations:

$$x + y = N , \tag{3.5}$$
$$3x + 5y = 210 . \tag{3.6}$$

(a) Here, N is given to be 46. The solutions for the two equations can be found by either working on the simultaneous equations directly, or by framing the two equations in matrix form. For the first method, we substitute $y = 46 - x$ from (3.5) into (3.6), giving $x = 10$ and $y = 36$. Using matrices, the two equations can be written as

$$\begin{pmatrix} 1 & 1 \\ 3 & 5 \end{pmatrix} \begin{pmatrix} x \\ y \end{pmatrix} = \begin{pmatrix} 46 \\ 210 \end{pmatrix}.$$

Hence,

$$\begin{pmatrix} x \\ y \end{pmatrix} = \frac{1}{2} \begin{pmatrix} 5 & -1 \\ -3 & 1 \end{pmatrix} \begin{pmatrix} 46 \\ 210 \end{pmatrix} = \begin{pmatrix} 10 \\ 36 \end{pmatrix}.$$

(b) For this, we look again at Equation (3.6) which says that $y = 42 - \dfrac{3x}{5}$. Since y should be an integer, therefore $x = 5k$ for some integer $k > 0$. Thus, $y = 42 - 3k$. We need to minimize the value of $|x - y| = |-42 + 8k|$, leading us to $k = 5$, that is $x = 25$. Hence, $y = 27$ and $N = 52$.

13. The total number of passengers will be 245. One bus from company X will be able to hold 40 passengers, compared to 35 for a bus from company Y. Let the number of buses from company X be x and those from company Y to be y. Therefore, $40x + 35y \geq 245$. The price for a bus from company X is three times of that from company Y, so $3x = y$. Solving for the two variables, we obtain $x = 1.8 \approx 2$, and therefore, $y = 6$.

14. (a) The matrices are as displayed in the question, that is

$$J = \begin{pmatrix} 5 & 1 & 3 & 2 & 4 \\ 4 & 3 & 5 & 2 & 5 \end{pmatrix}, \quad K = \begin{pmatrix} 4 & 2 & 3 & 3 & 3 \\ 2 & 3 & 4 & 3 & 5 \end{pmatrix}, \quad L = \begin{pmatrix} 3 & 4 & 2 & 4 & 3 \\ 1 & 3 & 3 & 4 & 4 \end{pmatrix}.$$

(b) The total score for each participant is the sum of all the scores from each of the judges. This operation can be represented by the addition of the matrices J, K and L. The matrix representing the total score T is therefore

$$T = J + K + L = \begin{pmatrix} 12 & 7 & 8 & 9 & 10 \\ 7 & 9 & 12 & 9 & 14 \end{pmatrix}.$$

(c) The weight matrix W needs to act on the matrix T such that the first row of matrix T (representing the total vocal scores) will be multiplied by 0.7, while the second row (measuring the originality) needs to be multiplied by 0.3. Therefore W can be written as $(0.7 \quad 0.3)$.

(d) The matrix consisting of the final score F is given by

$$WT = (0.7 \quad 0.3) \begin{pmatrix} 12 & 7 & 8 & 9 & 10 \\ 7 & 9 & 12 & 9 & 14 \end{pmatrix} = (10.5 \quad 7.6 \quad 9.2 \quad 9 \quad 11.2).$$

(e) The highest score, 11.2, is due to participant 'e'.

15. The prices are given by $C(x) = A + Bx$, with A and B being constants.

   (a) The prices for the 10 AU and 20 AU backpacks are \$70 and \$120 respectively. Hence,

$$70 = A + 10B,$$
$$120 = A + 20B.$$

   Solving the two equations gives us $B = 5$ and $A = 20$. When $x = 25$, we get $C = 20 + 5(5) = 145$.

   (b) For the waterproofed version of the backpack, the price is given by $S(x) = E + Fx + Dx^2$. From the information given, we can construct the following simultaneous equations:

$$80 = E + 10F + 100D,$$
$$111.25 = E + 15F + 225D,$$
$$145 = E + 20F + 400D.$$

   Solving these equations systematically, for example by first eliminating E between the first two and last two equations to get two equations in two variables, leads us eventually to $E = 25, F = 5$ and $D = 0.05$. Therefore, the price of a 25 UA waterproofed backpack will be $25 + 5(25) + 0.05(25)^2 = 181.25$ dollars.

   (c) A custom backpack of size x UA is priced according to the equation $P(x) = 50 + 5x + 0.05x^2$.

      i. If $P = 117.2$, then $0.05x^2 + 5x - 67.2 = 0$. Solving this quadratic equation, we obtain $x = 12$.

      ii. x represents the size of the backpack. Hence, $x \geq 0$. This means that in the equation $P(x) = 50 + 5x + 0.05x^2$, the second and third term on the right will always be zero or larger, so $P \geq 50$ always.

16. Let the rows and columns go in this order: Human, squid, fish and prawn. Based on given diagram, matrix A is indicated below. A^2 is a result of the multiplication of two A matrices, while A^3 is the multiplication of A^2 with the original matrix A.

$$A = \begin{pmatrix} 0 & 1 & 1 & 1 \\ 0 & 0 & 1 & 1 \\ 0 & 0 & 0 & 1 \\ 0 & 0 & 0 & 0 \end{pmatrix}, \quad A^2 = \begin{pmatrix} 0 & 0 & 1 & 2 \\ 0 & 0 & 0 & 1 \\ 0 & 0 & 0 & 0 \\ 0 & 0 & 0 & 0 \end{pmatrix}, \quad A^3 = \begin{pmatrix} 0 & 0 & 0 & 1 \\ 0 & 0 & 0 & 0 \\ 0 & 0 & 0 & 0 \\ 0 & 0 & 0 & 0 \end{pmatrix}.$$

Chapter 4

Geometry and Trigonometry

Exercises: 4.1 Angles and Lengths

1. $\sin(\theta) = 1.5/5$, so $\theta = \arcsin(1.5/5) = 0.38\,\text{radian} = 17.46°$.

   Note: arcsin() or $\sin^{-1}()$ *(and similarly for cosine and tangent) are usual conventions for the inverse of sine.*

2. The "hypotenuse", that is the distance travelled, is $2\text{m/s} \times 120\text{s} = 240\,\text{m}$. The constant angle of inclination is then $\theta = \arcsin(10/240) = 2.39°$.

3. (a) $x = 7\,\text{m} \times \tan(45°) = 7\,\text{m}$.

   (b) $x = 7\,\text{m}/\tan(55°) = 7/1.79\,\text{m} = 4.9\,\text{m}$.

4. (a) As shown in Fig.(4.1), the angle is the same. That is, we have $\tan(\theta) = 1/1.3 = x/5.5$, so $x = 5.5/1.3 = 4.23\,\text{m}$.

   (b) $\theta = \arctan(1/1.3) = 37.6°$.

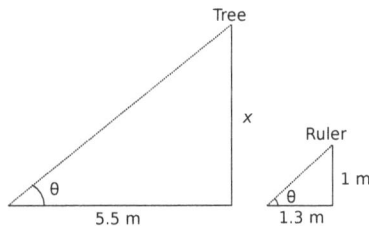

Figure 4.1: Similar Triangles.

5. Fig.(4.2) depicts the situation: The line of sight of Neo from the top of the tower to the horizon, is perpendicular to the line from the centre of the Earth to that horizon point, whose length is the radius of the Earth, R. (We take the Earth to be a sphere). Meanwhile, Neo is at a distance of $R + 50\,\text{m}$ away from the Earth's centre. The angle of depression is $0.23°$, which tells us that $\sin(89.77°) = R/(R + 50)$, so that $R \approx 6206\,\text{km}$.

6. Fig. (4.3) depicts the situation.

   (a) $x = 150\,\text{m} \times \tan(20°) = 54.6\,\text{m}$.

   (b) $y = x - 150\,\text{m} \times \tan(10°) = 28.15\,\text{m}$.

21

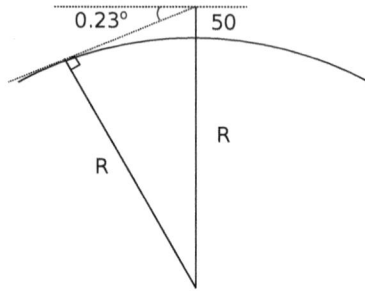

Figure 4.2: Schematic showing the angles and distances given.

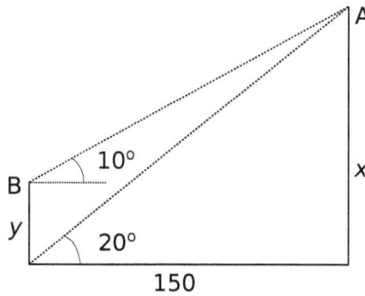

Figure 4.3: Schematic showing the relevant angles and distances.

7. For a circle of radius R in the x-y plane, we know that its y component is given by $y = R\cos\theta$ where θ is measured from the y axis. Now, the centre of the wheel is at a height of $h + R$, and thus the height of Eona above ground will be $h + R + y = h + R(1 + \cos\theta)$. See Fig. (4.4) for an illustration.

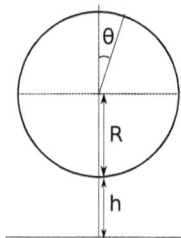

Figure 4.4: Schematic depicting the Ferris wheel.

8. (a) This is actually similar to the Exercise 6, where our task now is to find d, corresponding to the "distance between two buildings", given that the "angles of elevation" are $10°$ and $2°$ respectively, from the "ground" and "top" of the "building B", which has a "height" of $20\,\mathrm{km}$. Then, we have

$20 \text{ km} = d(\tan(10°) - \tan(2°)) = d \times 0.1414$, such that $d = 141.4 \text{ km}$.

(b) The distant star is like the mountain in problem 8(a), so by measuring the angles the star makes from different points, for example along the Earth's orbit around the Sun, one will be able to deduce the distance of the star from Earth.

9. (a) From Fig. (4.5), we see that $500 \text{ m} = \dfrac{x}{\tan(10°)} + \dfrac{x}{\tan(5°)}$, and thus $x = 29.2 \text{ m}$.

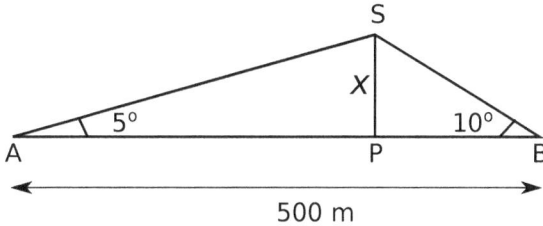

Figure 4.5: Schematic showing the relevant angles and distances.

(b) Assume that we know the distance from the Earth to the Moon. This plays the role of x in Fig. (4.5), with P as the center of the moon, and A to B measures the diameter of the Moon. Since P is at the center of the Moon, AP is equal to PB. By measuring the angle extended by the Moon (which is the angle of S of the triangle), we can then estimate the diameter of the Moon.

10. (a) From Fig. (4.6), we see that the angles of the triangle are $50°, 85°, 45°$. Then, by the sine rule, we have $\dfrac{200}{\sin(50°)} = \dfrac{AB}{\sin(45°)}$, so $AB = 184.6 \text{ m}$.

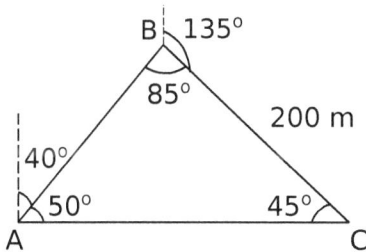

Figure 4.6: Schematic showing the angles and distances.

(b) $\dfrac{200}{\sin(50°)} = \dfrac{AC}{\sin(85°)}$ implies $AC = 260.1 \text{ m}$.

(c) We can use Area$= \frac{1}{2}ab\sin(\theta)$ where a and b are lengths of any two sides of the triangle, and θ is the angle subtended by these two sides. Then, $AB \times AC \times \sin(50°)/2 = 18,391 \text{ m}^2$.

11. (a) Refer to Fig.(4.7). Re-write Snell's law as $\frac{n_1}{n_2} = \frac{\sin(\theta_2)}{\sin(\theta_1)}$. Since $\theta_2 > \theta_1$, and both angles are smaller than $90°$, so $\sin(\theta_2) > \sin(\theta_1)$ and $\frac{n_1}{n_2} > 1$, that is $n_1 > n_2$.

(b) When $\theta_2 = \pi/2$, we have $\frac{n_1}{n_2} = \frac{1}{\sin(\theta_c)}$. Thus $\theta_c = \arcsin(n_2/n_1)$.

(c) First, note that θ_c is only meaningfully defined for $n_2 \leq n_1$, as otherwise $\arcsin(n_2/n_1)$ has no solution. Then, what 4.1.11(b) says is: When light is incident with an angle θ_c to the normal of the denser medium (with refractive index n_1), the light will exit the second medium (with refractive index n_2) by an angle of $\theta_2 = 90°$ to its normal, that is parallel to the surface of the second medium. Now, if the light is incident with an angle θ_1 larger than that of θ_c, we have $\sin(\theta_1) > \sin(\theta_c)$, and there will be no value of θ_2 such that $\frac{n_1}{n_2} = \frac{\sin(\theta_2)}{\sin(\theta_1)}$ holds, as we would need $\sin(\theta_2) > 1$. The conclusion is, there will be no refraction, that is, the light will not leave from the second medium. Rather, it will follow the usual *reflection* rule, as if the second medium is now a piece of mirror, reflecting the light back to the denser medium. This is called the *total internal reflection*. It has an application in fibre optics, where light is kept propagating inside the fibre by coating a less dense material around the core.

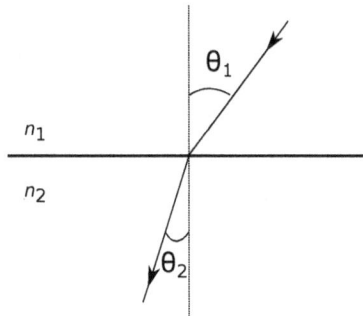

Figure 4.7: Schematic showing the rays and angles.

12. (a) $t = x/(U \cos \theta)$, and so $y = U \times \left(\dfrac{x}{U \cos \theta}\right) \times \sin \theta - \dfrac{g}{2} \times \left(\dfrac{x}{U \cos \theta}\right)^2 = x \tan \theta - (g/2U^2)x^2 \sec^2 \theta$.

(b) We want to find the value of x where $y = 0$. One solution is obviously 0, and another one will be the range R. It is given by $0 = \tan\theta - (g/2U^2)R\sec^2\theta$, that is $R = \tan\theta \times \dfrac{(2U^2/g)}{\sec^2\theta} = U^2\sin(2\theta)/g$.

(c) For a given U and constant g (which is positive), to maximize R we obviously need $\sin(2\theta) = 1$, that is, $\theta = 45°$, such that $R_{\max} = U^2/g$.

(d) With $\dfrac{d^2x}{dt^2} = 0$, integrate once to get $v_x(t) - v_x(0) = \displaystyle\int_0^t (0)\, dt = 0$, that is, $v_x(t)$ is a constant, and is equal to its initial value $v_x(0)$. The initial velocity is U, and hence its x component is $U\cos\theta$, that is, $v_x(t) = U\cos\theta$. Integrate the velocity to get the distance x, that is

$$x(t) = \int_0^t (U\cos\theta)\, dt = Ut\cos\theta.$$

Similarly, with $\dfrac{d^2y}{dt^2} = -g$, integrate once to get $v_y(t) - v_y(0) = \displaystyle\int_0^t (-g)\, dt = -gt$. $v_y(0)$ is $U\sin\theta$, and so $v_,y(t) = U\sin\theta - gt$. Integrate the velocity to get the distance y, that is $y(t) = \displaystyle\int_0^t (U\sin\theta - gt)\, dt = Ut\sin\theta - gt^2/2$.

(e) Intuitively, if air-resistance is included, $R_{\max}$ will be shortened. Details depend on the exact model of air-resistance used. To check that our intuition is correct, consider the example where air-resistance is proportional to $-v$ (see Chapter 2, Exercise 24). Then, we have

$$\frac{dv_x}{dt} = -bv_x,$$
$$\frac{dv_y}{dt} = -g - bv_y, \tag{4.1}$$

which gives us

$$v_x(t) = U\cos\theta\, e^{-bt},$$
$$v_y(t) = U\sin\theta\, e^{-bt} - \frac{g}{b}\left(1 - e^{-bt}\right), \tag{4.2}$$

and

$$x(t) = \frac{U\cos\theta}{b}\left(1 - e^{-bt}\right),$$
$$y(t) = \frac{U\sin\theta}{b}\left(1 - e^{-bt}\right) + \frac{g}{b^2}\left(1 - bt - e^{-bt}\right). \tag{4.3}$$

We leave it as an exercise for the reader to show that one recovers the previous no-air-resistance expressions by taking the limit $b \to 0$. (Hint: Use L'Hospital rule).

Now, for any fixed θ, the time taken to reach R for the no-air-resistance case is $T = T(\theta) = (2U \sin \theta)/g$. At this time, for the case of non-zero air-resistance with the same θ, the y-position will be $\left(\dfrac{U \sin \theta}{b} + \dfrac{g}{b^2}\right)\left(1 - e^{-2bU \sin \theta/g}\right) - \dfrac{2U \sin \theta}{b}$. Multiplying this expression by b^2/g, and defining $z \equiv bU \sin \theta/g$, we get $(z+1)(1-e^{-2z}) - 2z$. When $z = 0$, this expression is 0, and one can show that its gradient, which is equal to $e^{-2z}(1+2z)-1$, is always negative, since $1 + 2z < e^{2z}$ for all $z > 0$. That is, $y(T) < 0$ for any $b > 0$.

However, as the ground is at $y = 0$, this implies that the projectile will hit the ground at a time $t = T'(\theta) < T(\theta)$. Then, $x(T'(\theta)) < v_x(0)T'(\theta) = U \cos(\theta)T'(\theta) < U \cos(\theta)T = \dfrac{U^2 \sin(2\theta)}{g} \leq \dfrac{U^2}{g}$. That is,

$$\max_{\theta}\{x(T'(\theta))\} \equiv R'_{\max} < \frac{U^2}{g} \equiv R_{\max}. \tag{4.4}$$

In other words, with air-resistance, the range is decreased if the other variables are kept fixed.

Exercises: 4.2 Periodicity and Waves

1. (a) As θ can only range between 0 and $1/8$, one must have $A = 1/8$, since $\cos(\omega t + B)$ will explore all the values between 0 and 1. Then, when $t = 0$, $1/8 = 1/8 \times \cos B$, implies $B = 0$. The pendulum first returns to its original state after $t = T$, so $\cos(\omega T) = \cos(0) = 1 \Rightarrow \omega T = 2\pi$, so $\omega = 2\pi/T = 2\pi/3 \, \text{s}^{-1}$.

   (b) With $\cos x = \sin(x + \pi/2)$, $\theta = \dfrac{1}{8} \sin(2\pi t/3 + \pi/2)$.

2. (a) See Fig. (4.8).

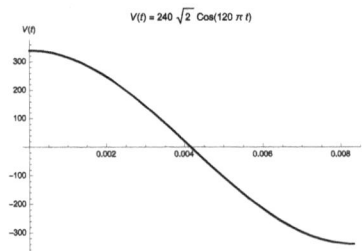

Figure 4.8: Plot of $V(t)$, with $V_0 = 240\sqrt{2}$ Volts and $f = 60\,\text{Hz}$.

(b) $V < 240 \Rightarrow \cos x < 1/\sqrt{2}$ where $x = 2\pi ft$. So $t \in (1/480, 7/480)$. With the constraint $t < 1/2f = 1/120$, we get $1/480 < t < 1/120$.

(c) We now need $\cos x > -1/4\sqrt{2}$. As $\arccos(-1/4\sqrt{2}) = 1.578$, inspection of the figure leads us to the solution $0 < t < 0.00464$.

(d) AC voltage can be stepped-up or down easily using transformers, with minimal energy loss.

3. (a) $y(L) = A\sin\dfrac{2\pi L}{\lambda} = 0 \Rightarrow 2\pi L/\lambda = n\pi$, that is $\lambda = 2L/n$ with n a positive integer.

(b) For $n = 2$, $y(x) = A\sin\dfrac{2\pi x}{L}$. The nodes occur when $y = 0$, that is at $x = 0$, $L/2$, L. The anti-nodes occur when $|y| = A$, that is at $x = L/4$, $3L/4$.

(c) "Harmonics" refers to frequencies of sound produced by the guitar when the string vibrates with $\lambda = 2L/n$. "Fundamental" refers to the case $n = 1$.

(d) Standing waves also occur in wave guides, which are structures used to constrain the propagation of electromagnetic waves.

4. (a) When $x = 0$, $y = A = 15$ m. At the maximum point, $y = 25 = 15 + B$, so $B = 10$ m. The two peaks differ by 2π in the argument of the sine, that is, $50k = 2\pi$, so $k = 2\pi/50\,\mathrm{m}^{-1}$.

(b) $y_{\min} = A - B = 5$ m (when $\sin kx = -1$).

5. (a) Recall that
$$A_2 \equiv \cos(A + B) = \cos A \cos B - \sin A \sin B,$$
$$A_1 \equiv \cos(A - B) = \cos A \cos B + \sin A \sin B. \tag{4.5}$$

Summing them, one gets $\cos(A+B) + \cos(A-B) = 2\cos A \cos B$. Identifying $A + B = 2\pi f_2 t$ and $A - B = 2\pi f_1 t$, we get $A(t) = A_1 + A_2 = 2\cos(\pi(f_1 + f_2)t)\cos(\pi(f_2 - f_1)t)$.

(b) See Fig. (4.9).

(c) As $f_1 \approx f_2$, $f_2 - f_1 \ll f_1 + f_2$, so the slowly-varying $2\cos(\pi(f_1 - f_2)t)$ will serve as an envelope for the fast oscillating $\cos(\pi(f_1 + f_2)t)$. This envelope will determine the amplitude of the fast oscillating sound wave $\cos(\pi(f_1 + f_2)t) = \cos(2\pi t(f_1 + f_2)/2)$, which has an average frequency of $(f_1 + f_2)/2$.

(d) Squaring the amplitude $2\cos(\pi(f_2 - f_1)t)$, we get $2(\cos(2\pi(f_2 - f_1)t) + 1)$, which is a periodic function with frequency $f_2 - f_1$. Since the loudness of the sound is proportional to the square of its wave amplitude, the above result tells us that one hears a variation of loudness at a frequency of $f_2 - f_1$, which is known as the *beat frequency*.

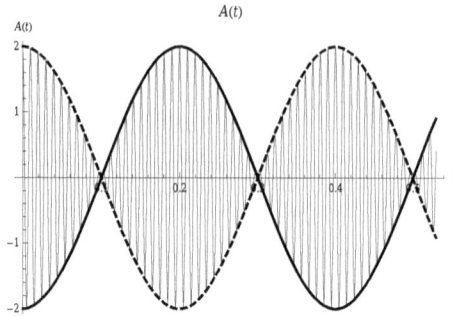

Figure 4.9: Plot of $A(t)$, with $f_1 = 100, f_2 = 105$. The thin line is $A(t)$, while the thick and dotted lines are the slow-varying envelope, that is $\pm 2\cos(\pi(f_2 - f_1)t)$.

(e) With the aid of a reference instrument, which is known to produce the wanted frequency f_1 precisely, the musical instrument will produce a beat if its frequency f_2 deviates slightly from f_1. One can then tune the musical instrument (adjust its frequency) until the beat slows down and is no longer detectable.

6. (a) First we re-write $A(x) = \sin\left(\dfrac{2\pi x}{5}\right) + 3\cos\left(\dfrac{2\pi x}{5}\right)$. Then using the R-formulae we get $A(x) = \sqrt{1+9}\sin(2\pi x/5 + \alpha)$ where $\alpha = \arctan 3 = 1.2490$. Thus, $A(x_{\max}) = \sqrt{10}$.

   (b) $2\pi x/5 + \alpha = \pi/2 \Rightarrow x = 0.256$.

   (c) $2\pi x/5 + \alpha = \pi \Rightarrow x = 1.506$.

7. (a) The maximum and minimum temperatures occur when $\sin(\omega t) = \pm 1$ respectively. So $A+B = 32$ and $A-B = 22$, giving $A = 27, B = 5$. Lastly, we have $26 = 27 + 5\sin(3\omega)$ which implies $3\omega = \arcsin(-1/5) = 3.343$, and $\omega = 1.114$. We have taken the value $\arcsin(-1/5) = 3.343$ instead of -0.201 because of the restriction $0 < \omega < \pi/2$.

   (b) $f(6) = 27 + 5\sin(6\omega) = 29°\mathrm{C}$.

8. (a) $y(L) = 0 \Rightarrow A\sin(kL) = 0 \Rightarrow k = n\pi/L$. Combining with $k = 2\pi/\lambda$, we get $\lambda = 2L/n$.

   (b) $E = \dfrac{\hbar^2 k^2}{2m} = \dfrac{\hbar^2 n^2 \pi^2}{2mL^2}$. Minimum energy is at $n = 1$, that is $E_{\min} = \dfrac{\hbar^2 \pi^2}{2mL^2}$.

   (c) Since E is proportional to n^2, the energy levels are not equally spaced with respect to n.

Chapter 5

Coordinate Geometry

1. The town is laid out in a rectangular grid, with the position of Eona given by (5,7) and the bookshop at (10,19). Each grid length corresponds to 50 m in length.

   (a) The bookshop is located $10 - 5 = 5$ grids horizontally and $19 - 7 = 12$ grids vertically away from Eona. This means that the distance that she has to walk if she follows the streets will be $(5 + 12) \times 50 = 850$ m.

   (b) The minimum distance will be a straight line between the two points. In terms of grids, the distance is given by Pythagoras' theorem, $\sqrt{5^2 + 12^2} = 13$, which corresponds to $13 \times 50 = 650$ m.

2. The points are A $= (10, 20)$, B $= (4, 26)$, C $= (19, 41)$, and D $= (25, 37)$, as shown in Fig. (5.1).

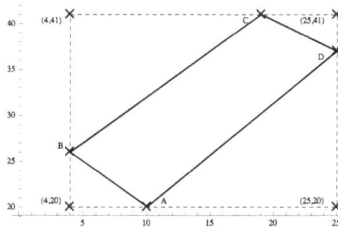

Figure 5.1: Figure showing the coordinates of the vertices.

   (a) Consider a square, with vertices given by $(4, 20), (4, 41), (25, 41)$, and $(25, 20)$. The total area of this square is $A_s = 21 \times 21 \times 100 \, \text{m}^2$. The area of the land chosen by the contractor can be computed as A_s minus the areas of 4 right-angled triangles. The total area of the 4 right-angled triangles is $27000 \, \text{m}^2$, so the answer is $17100 \, \text{m}^2$. (Another way is to divide the polygon into two triangles and use formula A.47 from the main text).

   (b) The plot chosen by the investor is not rectangular. This can be seen from Fig. (5.1). Alternatively, draw a line joining the points A and B, and another line joining the points C and D. The first line has a gradient of -1, while the second line has a gradient of -2/3. This means that the two lines are not parallel to each other, and hence, the plot cannot be rectangular.

   (c) The point to be shifted is the point D$(25, 37)$. The new coordinate (let us call it E) can be found by making sure that the line joining this point to C has the same gradient as the line joining A and B, that is, gradient of -1. Therefore, $y = -x + c$, where c is the y-intercept. Substituting the coordinates C(19,41) into the equation, we find that $c = 60$, that is line CD follows the equation $y = -x + 60$.

Since a rectangle is made up of parallel sides, the line joining B and C should be parallel to the line joining A and E, that is have a gradient of 1. Using the coordinates for A, we will find that the equation describing the line AD is $y = x + 10$.

Now, the intersection between lines CD and lines AD will produce the coordinate for $E(25, 35)$.

(d) The area of the rectangle will be the product of the lengths of AE and AB, both of which can be found using the formula $l = \sqrt{(x_1 - x_2)^2 + (y_1 - y_2)^2}$. That is, $l_{AE} = 15\sqrt{2}$ units while $l_{AB} = 6\sqrt{2}$ units. Therefore, the resulting area is 180 unit^2. Recall that 1 unit on the map corresponds to 10 m, so the area of the plot is 18000 m^2.

3. The distance travelled by the car is given by $x = t + t^2$.

(a) The graph of x versus t is as shown.

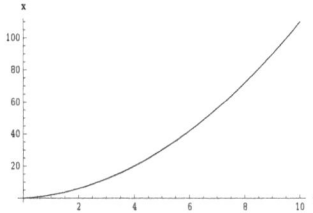

Figure 5.2: Graph of x versus t.

(b) The graph of $x/t = 1 + t$ against t is displayed in Fig. (5.3).

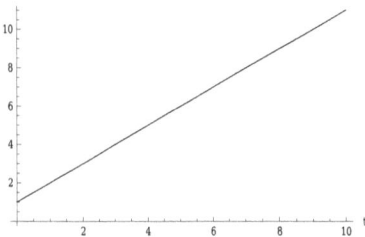

Figure 5.3: Plot of x/t versus t.

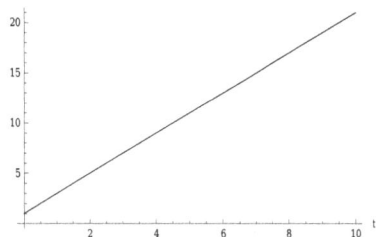

Figure 5.4: Graph of dx/dt versus t.

(c) $dx/dt = 1 + 2t$. The graph of dx/dt versus t is as shown in Fig. (5.4).

(d) The graphs exhibiting straight lines are graphs in (b) and (c), with gradients 1 and 2, respectively.

4. The equation describing the number of radioactive nuclei remaining after a time t is given by $N(t) = N_0 e^{-\lambda t}$.

(a) From the given equation, $\ln N(t) = -\lambda t + \ln N_0$. The graph of $y \equiv \ln N(t)$ is shown in Fig. (5.5).

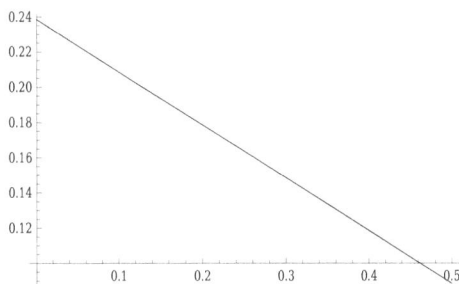

Figure 5.5: Plot of $\ln N(t)$ against t, using $N_0 = 100$, and $\lambda = 0.3$.

(b) From the equation $\ln N = -\lambda t + \ln N_0$, we see that in a plot of $y \equiv \ln N$ against t, the gradient will be $-\lambda$ and the y-intercept will be $\ln N_0$.

(c) Since $N(t)/N_0 = e^{-\lambda t}$, $\ln[N(t)/N_0] = -\lambda t$. So a plot of $\ln[N(t)/N_0]$ against t will result in a straight line with zero y-intercept.

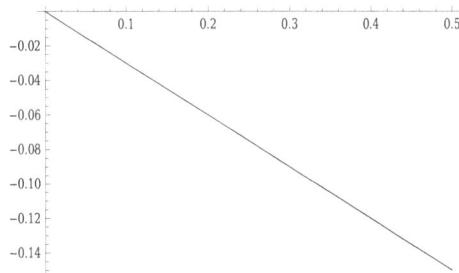

Figure 5.6: Graph of $\ln(N(t)/N_0)$ against t.

5. The energy is given by $E = \frac{1}{2}kx^2$.

(a) In a plot of E against x^2, the slope will be $k/2$ and the intercept will be at the origin.

(b) We get $\ln E = \ln(k/2) + 2\ln x$. Therefore, in a $\ln E$ against $\ln x$ plot, the gradient will be 2 and the y-intercept will be $\ln(k/2)$.

6. We have $d = kv^2$, where k is a constant. For a straight line plot involving d and v, we can either

   (a) Plot d against v^2, such that the plot will have a gradient of k, or

   (b) Plot $\log d$ against $\log v$. In this case, $\log d = \log k + 2 \log v$, so the gradient will be 2.

7. The cost to produce x number of waffles is given by

$$C(x) = \frac{60x + 300}{0.1x + 3} = \frac{600x + 3000}{x + 30} \,. \tag{5.1}$$

   (a) Therefore

$$C(x) = \frac{600(x + 30) + 3000 - 18000}{x + 30} \tag{5.2}$$

$$= 600 - \frac{15000}{x + 30} \,. \tag{5.3}$$

   i. As $x \to \infty$, the term $\dfrac{15000}{x + 30} \to 0$. Hence, $C(x) \to 600$.

   ii. The graph of $C(x)$ against x is shown in Fig. (5.7).

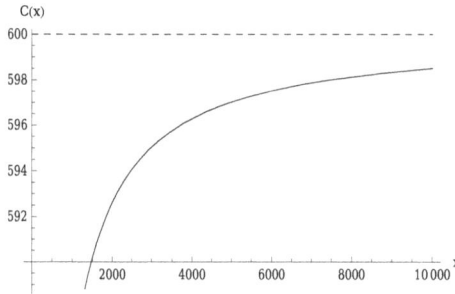

Figure 5.7: Plot of $C(x)$ against x.

   iii. As x increases, the term $\dfrac{15000}{x + 30}$ decreases, which means that $C(x)$ will increase, up to the asymptote at $C = 600$.

   (b) The profit is given by $P(x) = R(x) - C(x)$, where $R(x) = 3x$. For $x = 100$, we obtain $P(100) < 0$, which indicates a loss. On the other hand, $x = 200$ gives $P(200) > 0$, which means a profit.

   (c) Plotting $C(x)$ and $R(x)$ on the same graph will allow us to see the break-even point, see Fig. (5.8). The break-even point is around $x = 175$.

   (d) The break-even point can be found by equating the revenue R with the cost C. Therefore, $3x = 600 - 15000/(x + 30)$. This gives $x^2 - 170x - 1000 = 0$, that is $x = 175.7$.

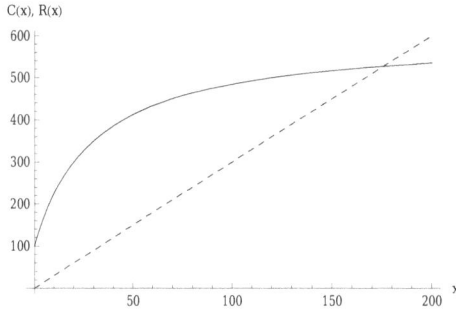

Figure 5.8: Graph of $C(x)$ and $R(x)$ (straight line) against x.

8. The data relates distance a with orbital period T.

    (a) For Earth the values of both a and T are 1. This means that the units for a and T are measured relative to Earth.

    (b) The plot of $\log T$ against $\log a$ is shown in Fig. (5.9).

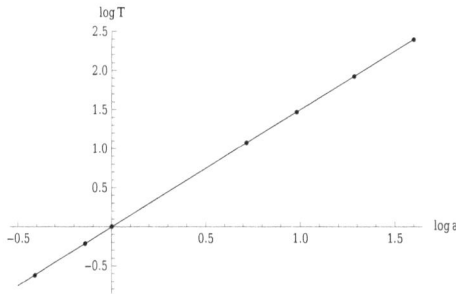

Figure 5.9: Plot of $\log T$ against $\log a$.

    (c) In (b), the best fit equation (using the least-squares method) is given by $\log(T) = -0.002 + 1.502 \log(a)$. This tells us that $\log(T) \approx 1.5 \log(a)$, that is $T = a^{1.5}$. (The base of the logarithm used does not matter).

    (d) As $T = a^{1.5}$, therefore, $p = 1.88^{2/3} = 1.52$ and $q = 30.06^{1.5} = 165$.

    (e) Here, $T = 2.7$. Therefore, $a = (2.7)^{2/3} = 1.94$.

    (f) As $a = 25$, so from Kepler's law, $T = (25)^{1.5} = 125$.

9. The relation is $I = Ae^{-\mu d}$.

    (a) We obtain $\ln(I/A) = -\mu d$. Since μ is a constant, the graph of $\ln I/A$ against d will be a straight line with slope $-\mu$, see Fig. (5.10).

    (b) Lead would be a better material, as it shows a greater decrease in the I/A value with distance.

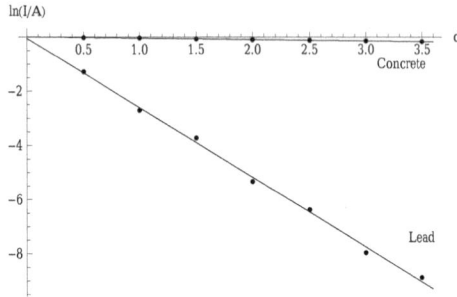

Figure 5.10: Graphs of $\ln(I/A)$ against d. The dots correspond to empirical data, while the lines are their respective best linear fits.

(c) You can draw a line of best fit by inspection. We leave it to you.

(d) Using freely available software that implements the least-squares method, the best linear fit for concrete is given by the formula $\ln(I/A) = 0.0031 - 0.0477d \approx -0.0477d$. The best linear fit for lead is given by $\ln(I/A) = -0.048 - 2.566d \approx -2.57d$. Therefore, the attenuation coefficients are lead $\approx 2.57\,\text{mm}^{-1}$, concrete $\approx 0.0477\,\text{mm}^{-1}$.

(e) From part (d) we deduce that 2 mm of lead will result in the $-\ln(I/A)$ value $2(2.57)$, which is to be equated to the value $0.0477d$ for concrete. Hence, $d = 2(2.57)/0.0477 \approx 108$ mm.

(f) First, the intensity of the x-ray after passing through 0.5 mm of lead is calculated to be $0.277A$. Then when that radiation passes through 50 mm of concrete, the emerging intensity is $I = 0.277Ae^{-0.0477\times 50} = 0.025A$.

(g) Hospital X-ray rooms.

10. The positions are described by $x_P = 2\sin(3t)$ and $x_P = 2\sin(6t)$.

(a) The graphs of the functions x_P and x_Q are shown below.

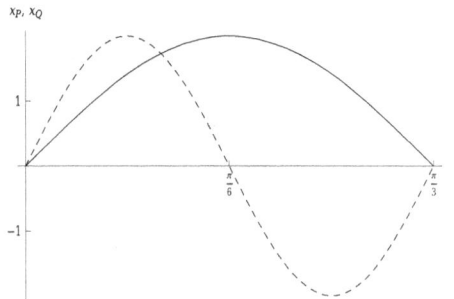

Figure 5.11: Graphs of x_P (solid line) and x_Q against t for $0 \leq t \leq \pi/3$.

(b) From the graph, the two curves intersect at $t = 0, \pi/9$, and $\pi/3$.

(c) At the intersection, $x_P = x_Q$, that is $\sin(3t) = \sin(6t)$. Using the trigono-metric identity $\sin(2A) = 2\sin(A)\cos(A)$, we deduce that $2\cos(3t) = 1$ or $\sin(3t) = 0$. This means that t can take the values of $0, \pi/3$, or $\pi/9$ in the range $0 \le t \le \pi/3$.

11. The length of fencing material available is 100 m. The area to be fenced up forms a rectangle with one side bordering a river.

(a) Let the side parallel to the river be of length l and the remaining two sides, which are perpendicular to the river, be x. Therefore, the total area A bounded by the fence will be $A = xl$. Since the total length of the fence is $100 = l + 2x$, we can write $A = x(100 - 2x)$.

(b) The graph of A as a function of x is shown below.

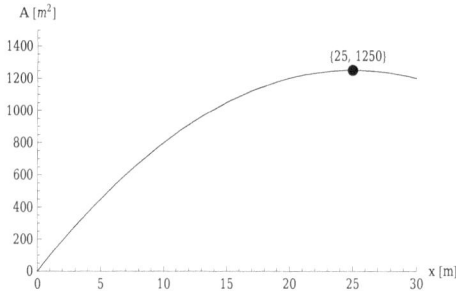

Figure 5.12: A as a function of x. The maximum is at $x = 25$ m and $A = 1250$ m^2.

(c) Two methods can be used to find the maximum of $A = x(100 - 2x)$.

i. Completing the square in $A(x) = 100x - 2x^2$:

$$x^2 - 50x = -\frac{A}{2}$$

$$x^2 - 50x + 625 = -\frac{A}{2} + 625$$

$$(x - 25)^2 = -\frac{A}{2} + 625$$

$$A = -2(x - 25)^2 + 1250.$$

Since $(x - 25)^2 \ge 0$, the maximum value of A is obtained when the equal sign holds, that is $x = 25$, giving $A_{max} = 1250$ m^2.

ii. Using calculus, extremum values satisfy $0 = dA/dx = 100 - 4x \Rightarrow$ $x = 25$ m and $A_{max} = 1250$. To ensure that a maximum occurs at

$x = 25\,\mathrm{m}$, we need to check that $d^2A/dx^2 < 0$, which is the case here. (Since the smooth function $A(x)$ vanishes at the boundaries $x = 0$ and $x = 50$, the local maximum found here must be the global maximum.)

(d) We suppose that the length of the fence is now L. This means that $L = l + 2x$, and $A = x(L - 2x)$. Using calculus, $0 = dA/dx = L - 4x \Rightarrow x = L/4$. Substituting this expression into $A = x(L - 2x)$, we obtain $A = L^2/8$. (Please check that this corresponds to maximum A).

12. The revenue $R(x)$ is dependent upon the tax rate x, $(0 \le x \le 100)$.

(a) The graph of $R(x) = \dfrac{1}{200}(x^3 - 300x^2 + 20000x)$ is shown in Fig. (5.13) while that of $R(x) = \dfrac{400x - 4x^2}{5x + 150}$ is shown in Fig. (5.14). The maxima are indicated.

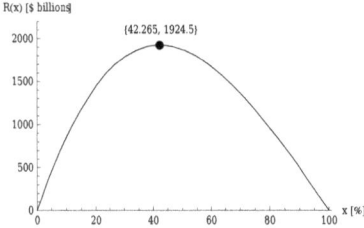

Figure 5.13

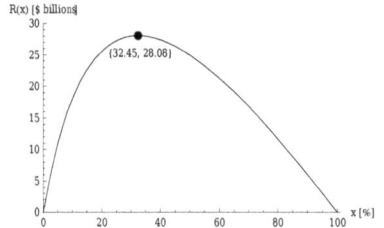

Figure 5.14

(b) The extrema are located at $dR/dx = 0$. In the first case,

$$dR/dx = \frac{1}{200}(3x^2 - 600x + 20000) = 0 \Rightarrow x = 42.265\%,$$

while in the second case,

$$dR/dx = \frac{(5x + 150)(400 - 8x) - 5(400x - 4x^2)}{(5x + 150)^2} = 0 \Rightarrow x = 32.4\%.$$

These values correspond to locations of global maxima, as seen from the plots in part (a).

Chapter 6

Differential Calculus

1. (a) From Q7 of Chap. 4.1, we have $H = h+R(1+\cos(\theta)) = h+R(1+\cos(\omega t))$. Hence, $\dfrac{dH}{dt} \equiv v = -\omega R\sin(\omega t)$.

   (b) The fastest height change means the highest speed v. At such positions, the magnitude of acceleration, that is the time derivative of the speed, is zero, that is $a = \dfrac{dv}{dt} = -\omega^2 R\cos(\omega t) = 0$, which is satisfied for $\omega t = \theta = \pi/2$ or $3\pi/2$.

2. From Q16 of Chap. 1, we have $V = x(a-2x)^2 \leq V_{max} = 2a^3/27$. So we need to set $V_{max} \geq 1000$, which implies $a \geq 15(4)^{1/3}$ cm.

3. (a) $v' = \dfrac{d}{dt'}(x - ut)$. But since $t = t'$, we can write $v' = \dfrac{d}{dt}(x - ut) = v - u$, where $v = \dfrac{dx}{dt}$. (Note: u is a constant).

   (b) As above, $a' = \dfrac{d}{dt'}v' = \dfrac{d}{dt}(v - u) = a$, where $a = \dfrac{dv}{dt}$.

   (c) The result is as expected in Newtonian physics for two frames moving at uniform relative velocity.

4. With $v \equiv \dfrac{dx}{dt}$,

   (a) $\dfrac{dx'}{dt} = \dfrac{d}{dt}\gamma(x - ut) = \gamma(v - u)$.

   (b) $\dfrac{dt'}{dt} = \dfrac{d}{dt}\gamma\left(t - \dfrac{ux}{c^2}\right) = \gamma\left(1 - \dfrac{uv}{c^2}\right)$.

   (c) By the chain rule, $v' = \dfrac{dx'}{dt'} = \dfrac{dx'}{dt}\dfrac{dt}{dt'} = \dfrac{dx'}{dt}\left(\dfrac{dt'}{dt}\right)^{-1} = \dfrac{v - u}{1 - \dfrac{uv}{c^2}}$.

   (d) $a' = \dfrac{dv'}{dt'} = \dfrac{dv'}{dt}\dfrac{dt}{dt'} = \dfrac{dv'}{dt}\left(\dfrac{dt'}{dt}\right)^{-1}$. From part (b), we have $\left(\dfrac{dt'}{dt}\right)^{-1} = \dfrac{1}{\gamma\left(1 - \dfrac{uv}{c^2}\right)}$, while from part (c) we have

$$\dfrac{dv'}{dt} = \dfrac{a}{1 - \dfrac{uv}{c^2}} - \dfrac{v - u}{\left(1 - \dfrac{uv}{c^2}\right)^2}\left(\dfrac{-ua}{c^2}\right)$$

$$= a\left(\dfrac{1}{1 - \dfrac{uv}{c^2}}\right)^2\left(1 - \dfrac{u^2}{c^2}\right),$$

37

where $a = \dfrac{dv}{dt}$. Therefore, $a' = \dfrac{a}{\gamma}\left(\dfrac{1}{1 - \dfrac{uv}{c^2}}\right)^3\left(1 - \dfrac{u^2}{c^2}\right)$.

5. It is given that $N(x) = -x^3 + 2x^2 + 300x + 4000$, where x represents the temperature of the culture in °C. Hence, $\dfrac{dN}{dx} = -3x^2 + 4x + 300$.

(a) Local extrema have $\dfrac{dN}{dx} = 0$, that is $x = \dfrac{-4 - \sqrt{3600 + 16}}{-6} = 10.7°C$ in the indicated range.

(b) That this is indeed a local maximum can be verified by checking for $\dfrac{d^2 N}{dx^2} < 0$. With $\dfrac{d^2 N}{dx^2} = -6x + 4$, we indeed have $-6(10.7) + 4 = -60.2 < 0$. A sketch confirms it to be a global maximum in the indicated range.

6. (a) $v \equiv \dfrac{dS}{dt} = u - gt$.

(b) $a = \dfrac{dv}{dt} = -g$.

(c) Maximum height is reached when it momentarily comes to a stop, that is, when $v = 0$. This occurs at $t = u/g$. The maximum height is then
$$S_{\max} = S_0 + \dfrac{u^2}{2g}.$$

(d) $S = \dfrac{gt^2}{2} - ut - S_0 = 0$ when the ball hits the ground, that is $t = \dfrac{u + \sqrt{u^2 + 2gS_0}}{g}$.

7. (a) With $\dfrac{d}{dx}\sin(kx) = k\cos(kx)$ and $\dfrac{d}{dx}\cos(kx) = -k\sin(kx)$, we get
$\dfrac{d^2}{dx^2}\sin(kx) = -k^2\sin(kx)$, and $\dfrac{d^2}{dx^2}\psi(x) = -\left(\dfrac{n\pi}{L}\right)^2\psi(x)$.

(b) Multiply the result from part (a) by $-\dfrac{\hbar^2}{2m}$, to get $-\dfrac{\hbar^2}{2m}\dfrac{d^2}{dx^2}\psi(x) = \dfrac{\hbar^2}{2m}\left(\dfrac{n\pi}{L}\right)^2\psi(x)$, from which we can define $E_n \equiv \dfrac{\hbar^2}{2m}\left(\dfrac{n\pi}{L}\right)^2$, such that $-\dfrac{\hbar^2}{2m}\dfrac{d^2}{dx^2}\psi(x) = E_n\psi(x)$.

8. (a) A sketch shows that $C(x)$ has one turning point, a minimum, for $x > 0$.
Setting $0 = \dfrac{d}{dx}C(x) = 10 - \dfrac{50000}{(x + 1)^2}$ we get $(x + 1)^2 = 5000$, and so
$x = \sqrt{5000} - 1 = 69.711$.

(b) A sketch shows that $R(x)$ has one turning point, a mamimum, for $x > 0$.

Setting $0 = \dfrac{d}{dx} R(x) = 100 e^{-(x/100)^2} - \dfrac{x^2}{50} e^{-(x/100)^2}$, we get $100 - \dfrac{x^2}{50} = 0$.

Therefore, $x = 50\sqrt{2}$.

(c) From a plot of $P(x)$, see Fig. (6.1), we find the range for $P(x) \geq 0$ to be approximately $31.5 \leq x \leq 124.5$.

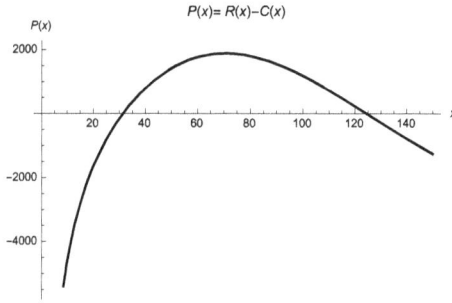

Figure 6.1: Plot of $P(x)$.

(d) From Fig. (6.1) or Fig. (6.2), we see that the maximum is around 70. More precisely, the value is $x = 70.571$.

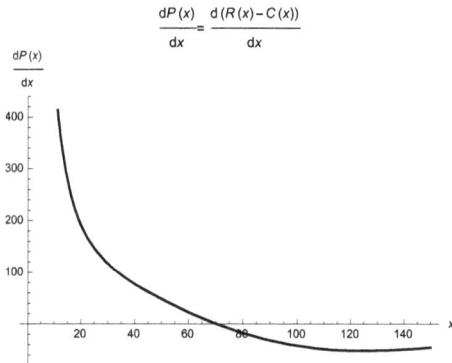

Figure 6.2: Plot of $\dfrac{dP(x)}{dx}$.

9. (a) We are given $\dfrac{dV}{dt} = 0.02 \, \text{m}^3\text{s}^{-1}$. For a sphere, $V = \dfrac{4}{3}\pi R^3$, we have

$\dfrac{dV}{dt} = 4\pi R^2 \dfrac{dR}{dt}$. Thus, for $R = 0.8 \, \text{m}$, we have $\dfrac{dR}{dt} = \dfrac{0.02}{4 \times \pi \times 0.64} = 0.0025 \, \text{ms}^{-1}$.

(b) For a sphere, $S = 4\pi R^2 \Rightarrow \dfrac{dS}{dt} = 8\pi R \dfrac{dR}{dt}$. Plugging in the values, we get $\dfrac{dS}{dt} = 0.05\,\mathrm{ms}^{-1}$.

(c) $\dfrac{dV}{dt} = 4\pi R^2 \dfrac{dR}{dt}$. Thus, $\dfrac{dR}{dt} = \dfrac{dV}{dt}\dfrac{1}{4\pi R^2}$. Putting this into $\dfrac{dS}{dt} = 8\pi R \dfrac{dR}{dt}$, we get

$$\frac{dS}{dt} = \frac{8\pi R}{4\pi R^2}\frac{dV}{dt} = \frac{2}{R}\frac{dV}{dt}, \tag{6.1}$$

that is $\dfrac{dS}{dt} \propto \dfrac{1}{R}$ when dV/dt is constant, as in this problem.

10. (a) As the top is square, set the width and breadth of the box to equal l, and let the height be h. The box then has volume $V = l^2 h$, and the decorated surface area is $S_c = l^2 + 4lh$. Then $V = l^2 h \Rightarrow l = \sqrt{\dfrac{V}{h}}$, and so $S_c = \dfrac{V}{h} + 4\sqrt{Vh}$. The volume V is fixed, and we want to minimize the surface area S_c. That is, we want $\dfrac{dS_c}{dh} = -\dfrac{V}{h^2} + 2\sqrt{\dfrac{V}{h}} = 0$, that is $h = \left(\dfrac{V}{4}\right)^{1/3}$, and $l = (2V)^{1/3}$. (A sketch of S_C confirms this solution to be a global minimum).

(b) The cost is $C \times S_c$, which for the minimal area is

$$C\times\left[(2V)^{2/3}+4\left(\frac{V}{4}\right)^{1/3}(2V)^{1/3}\right] = C\times\left[(2V)^{2/3}+2(2V)^{2/3}\right] = 3C(2V)^{2/3}.$$

11. (a) $x = 0 \Rightarrow \cos(2t) = 0 \Rightarrow 2t = \pi/2$ or $3\pi/2$, that is $t = \pi/4$ for the first return.

(b) $v = \dfrac{dx}{dt} = -10\sin(2t)$. At $t = \pi/4$, $\sin(2t) = \sin(\pi/2) = 1 \Rightarrow v = -10$.

(c) $v = -10\sin(2t) = 0 \Rightarrow 2t = \pi \Rightarrow t = \pi/2$.

(d) $a = \dfrac{dv}{dt} = -20\cos(2t)$. At $t = \pi/2$, $\cos(2t) = \cos(\pi) = -1 \Rightarrow a = 20$.

12. (a) We are given $\dfrac{dV}{dt} = aV^p - bV^q$ where a, b, p, q are positive constants. V represents the size of the tumour, and is therefore positive. Thus, both the terms aV^p and bV^q will be non-negative. $\dfrac{dV}{dt}$ is the rate of change in the tumour size, and hence the positive term aV^p represents growth, while the negative term $-bV^q$ is for degradation.

(b) We want to show that for $q > p$, there is always a maximum size to the tumour. We first re-write $\dfrac{dV}{dt} = aV^p\left[1 - \left(\dfrac{b}{a}\right)V^{q-p}\right]$. Local extrema occur when the derivative vanishes, that is when $V = 0$ or $V \equiv V_c = \left(\dfrac{a}{b}\right)^{1/(q-p)}$.

We first note that as $V \to 0$, $\dot{V} \approx aV^p$ as it will be dominated by the smaller power of V. Since $V \geq 0$, this implies that V increases as it deviates from zero. Thus $V = 0$ is a minimum point.

Next we re-write the derivative again as $\dfrac{dV}{dt} = aV^p\left[1 - \left(\dfrac{V}{V_c}\right)^{q-p}\right]$. From this we see that the term in square-brackets is positive for $V < V_c$ and negative for $V > V_c$. Thus we conclude that for $V > 0$, the tumour increases in size up to a maximum of V_c.

13. $\dfrac{dx}{dt} = Ax(B - x)$, A and B are positive constants.

(a) By simple observation, $x = 0$ and $x = B$ satisfy $\dfrac{dx}{dt} = 0$.

(b) When $x > B$, $Ax(B - x) < 0$, and so $\dot{x} < 0$. This means the population decreases.

(c) By the product rule, we have

$$\frac{d}{dt}\frac{dx}{dt} = A\frac{dx}{dt}(B - x) + Ax(-\frac{dx}{dt})$$
$$= A\dot{x}(B - x) - Ax\dot{x} = A\dot{x}(B - 2x). \tag{6.2}$$

(d) We have $\dfrac{dt}{dx} = \left(\dfrac{dx}{dt}\right)^{-1} = \dfrac{1}{Ax(B - x)} = \dfrac{1}{AB}\left(\dfrac{1}{x} + \dfrac{1}{B - x}\right)$. Integrating, this gives us $t = \dfrac{1}{AB}\ln\left(\dfrac{kx}{B - x}\right)$, after recognizing the anti-derivative of $1/x$ as $\ln x$. k here is a constant of integration. Inverting the expression, we get $x(t) = \dfrac{Be^{ABt}}{k + e^{ABt}}$. One could verify that this expression does indeed satisfy the logistic equation. See Fig. (6.3) for the curve of $x(t)$.

(e) We know B is one of the roots for the logistic equation. Also, by taking the limit $t \to \infty$, $x(t) \to B$. That is, B is the the steady-state solution, or in this case, the stable limiting population.

14. $\dfrac{dx}{dt} = x(a - bx)$, a and b are positive constants. Let $x(t = 0) = x_0$.

(a) $\dfrac{dx}{dt}\Big|_{t=0} = x_0(a - bx_0) = ax_0 - bx_0^2$. If we assume that x is actually the density (number per unit area) of infected animals, and that the initial

$$x(t) = \frac{Be^{ABt}}{k + e^{ABt}}$$

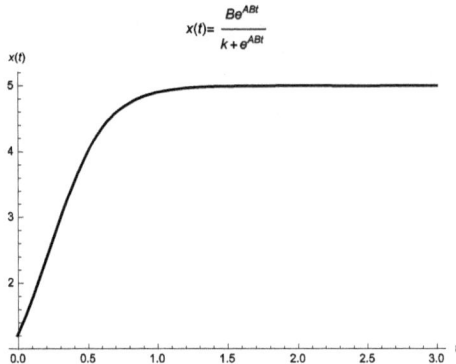

Figure 6.3: Plot of $x(t)$, with $A = 1, B = 5, k = 3$.

density, x_0, is small (< 1) then $\dot{x}(0) \approx ax_0$ as the smaller power of x_0 will dominate. So we deduce $a \approx 3$.

(b) The logistic equation in this exercise is the same as the previous exercise, with the conversion $A = b$, and $AB = a$. Hence, the solution of this exercise is $x(t) = \dfrac{\frac{a}{b}e^{at}}{k + e^{at}}$. Taking the limit $t \to \infty$, we get $a/b = 1000$, or $b = a/1000$. For $a \approx 3$, we have $b \approx 0.003$.

15. For a sphere of radius r, its volume is $V = 4\pi r^3/3$, and its surface area is $S = 4\pi r^2$. we will use Eq. (A.59) of the main text.

   (a) $\Delta V \approx \dfrac{dV}{dr}\Delta r \Rightarrow \Delta V \approx (4\pi r^2)\Delta r$ or $\Delta r = \Delta V/(4\pi r^2)$. Also, $r = (3V/4\pi)^{1/3} = (300/4\pi)^{1/3}$. Thus we get $\Delta r \approx 0.009598\,\text{cm}$.

   (b) With $\dfrac{dS}{dr} = 8\pi r$, we get $\Delta S \approx \Delta r \times 8\pi r \approx 0.6946\,\text{cm}^2$.

   (c) The exact radius at volume V is $r = (3V/4\pi)^{1/3}$. The exact difference in radius is therefore $(303/4\pi)^{1/3} - (300/4\pi)^{1/3} \approx 0.009566\,\text{cm}$. The inaccuracy in part (a) is thus about 10^{-5}.
   Similarly, the exact difference in surface area is approximately 0.6934. The inaccuracy in ΔS is then about $10^{-3}\,\text{cm}^2$.

16. $x(t) = 2e^{-2t}\cos(6t)$.

   (a) Since e^{-2t} is never zero for any finite t, $x(t) = 0$ implies $\cos(6t) = 0$. The smallest solution is $6t = \pi/2$, that is $t = \pi/12 \approx 0.26$.

   (b) $v(t) = \dfrac{dx}{dt} = -4e^{-2t}\cos(6t) - 12e^{-2t}\sin(6t)$. Then, $v(\pi/12) = -12e^{-\pi/6}$.

   (c) $v(t) = 0 \Rightarrow \cos(6t) + 3\sin(6t) = 0$. That is, $\tan(6t) = -1/3$, $\Rightarrow 6t = 2.82$, or $t = 0.47$.

(d) $a(t) = \dfrac{dv}{dt} = 8e^{-2t}\cos(6t) + 48e^{-2t}\sin(6t) - 72e^{-2t}\cos(6t)$. So $a(t) = 0 \Rightarrow -4\cos(6t) + 3\sin(6t) = 0$. That is, $\tan(6t) = 4/3$, or $t = 0.15$.

(e) See Fig. (6.4).

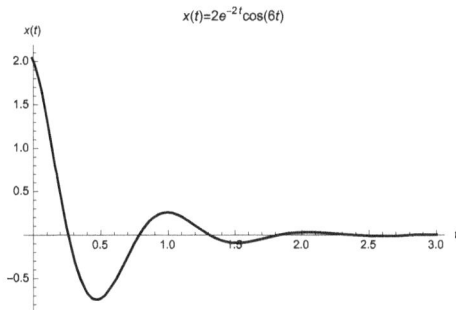

$$x(t) = 2e^{-2t}\cos(6t)$$

Figure 6.4: Plot of $x(t)$.

17. (a) For $x(t) = A\sin(at)$, $\dot{x} = Aa\cos(at)$, $\ddot{x} = -Aa^2\sin(at) = -a^2x(t)$. Thus, $\ddot{x} + a^2x = 0$ is indeed satisfied.

(b) For $x(t) = B\sin(bt)$, $\ddot{x} = -Bb^2\sin(bt) = -b^2x$. Then, $\ddot{x} + a^2x = (-b^2 + a^2)B\sin(bt) = F\sin(bt)$. Therefore, $B = F/(a^2 - b^2)$.

(c) As b approaches a, $a^2 - b^2$ gets smaller. Hence, B increases.

(d) In reality, there will be damping in the oscillator system, and hence the equation would not be exactly as given above. The solution would be different, with a bound for the amplitude (See Q22).

(e) Selecting a radio channel by tuning the resonance frequency.

18. (a) $\dfrac{dN(t)}{dt} = -\lambda N(t)$ has the well-known exponential solution $N(t) = Ae^{-\lambda t}$. The constant A is determined by $N(t_0) = Ae^{-\lambda t_0} = N_0$, that is $A = N_0 e^{\lambda t_0}$. Thus, $N(t) = N_0 e^{-\lambda(t-t_0)}$.

(b) $\left(\dfrac{dN_Y}{dt} + \lambda_2\right)N_Y = \lambda_1 N_X e^{-\lambda_1 t}$. The right-hand side is an exponential, for which we know its anti-derivative: An exponential itself. When the right-hand side is zero, we also know that the solution is an exponential. Hence, the solution is in the form $N_Y(t) = Ae^{-\lambda_2 t} + Be^{-\lambda_1 t}$, and our task is to determine the constants A, B. Using this trial expression in the given equation, we have

$$\dfrac{dN_Y}{dt} + \lambda_2 N_Y$$
$$= -A\lambda_2 e^{-\lambda_2 t} - B\lambda_1 e^{-\lambda_1 t} + \lambda_2(Ae^{-\lambda_2 t} + Be^{-\lambda_1 t})$$
$$= B(\lambda_2 - \lambda_1)e^{-\lambda_1 t}. \qquad (6.3)$$

Thus, $B = \dfrac{\lambda_1 N_X}{\lambda_2 - \lambda_1}$. A is determined by the requirement $N_Y(t = 0) = 0$. That is, $A = -B$.

19. We rewrite the equation as $\dfrac{d^2\psi}{dx^2} = -\dfrac{2mE}{\hbar^2}\psi \equiv \gamma^2\psi$. The general solution for this equation is $\psi(x) = A\cos(\gamma x) + B\sin(\gamma x)$. The boundary condition $\psi(0) = 0$ implies $A = 0$. The boundary condition $\psi(L) = B\sin(\gamma L) = 0$ is true if $B = 0$, but that implies $\psi \equiv 0$, which is a trivial solution. Non-trivial solutions are however possible: When $\gamma L = n\pi$, that is $E_n = \dfrac{n^2\hbar^2\pi^2}{2mL^2}$, $\psi(L) = 0$ is satisfied with B not being zero. We determine B by requiring $\displaystyle\int_0^L dx\, B^2 \sin\left(\dfrac{n\pi}{L}x\right)^2 =$
1. With $\displaystyle\int_0^L dx \sin\left(\dfrac{n\pi}{L}x\right)^2 = \int_0^L dx\, \dfrac{1 - \cos(n x\pi/L)}{2} = L/2$, we have $B = \sqrt{2/L}$. Therefore, $\psi(x) = \sqrt{\dfrac{2}{L}}\sin\left(\dfrac{n\pi}{L}x\right)$, as given in Eq. (6.1) of Q7.

20. We have $\dfrac{dt}{dx} = \left(\dfrac{dx}{dt}\right)^{-1} = \dfrac{1}{k(1-x)(2-x)} = \dfrac{1}{k}\left(\dfrac{1}{1-x} - \dfrac{1}{2-x}\right)$. Then, $t = \dfrac{1}{k}\ln\left(\gamma\dfrac{2-x}{1-x}\right)$, after recognizing the anti-derivative of $1/x$ as $\ln x$. γ is an integration constant. The initial condition $x = 0$ at $t = 0$ implies $\ln(2\gamma) = \ln 1 = 0$, that is $\gamma = 1/2$. So, $kt = \ln\left(\dfrac{2-x}{2-2x}\right)$, and exponentiating both sides, we have $e^{kt} = \dfrac{2-x}{2-2x}$, such that $(2-2x)e^{kt} = (2-x)$, or $x = \dfrac{2(e^{kt}-1)}{2e^{kt}-1}$.

21. (a) Completing the square in the numerator, we have
$$-\dfrac{dR}{dx} = -\dfrac{(x+50)^2 - 7500}{(x+50)^2} = 1 - \dfrac{7500}{(x+50)^2}.$$ This is now easily integrated with the given initial condition to yield $R = \dfrac{x(100-x)}{x+50}$.

(b) When $x = 100$, $R = \dfrac{100(100-100)}{100+50} = 0$.

(c) See Fig. (6.5).

22. (a) By repeated use of the product rule, it is straightforward to obtain
$$\dfrac{dy}{dt} = -dy - \omega B(t),$$
$$\dfrac{d^2y}{dt^2} = (d^2 - \omega^2)y + 2\omega dB(t), \tag{6.4}$$

where $B(t) = Ae^{-dt}\sin(\omega t + \phi)$. Thus, we require that
$m(d^2 - \omega^2)y + 2m\omega dB(t) \equiv -c\left(-dy - \omega B(t)\right) - ky$.

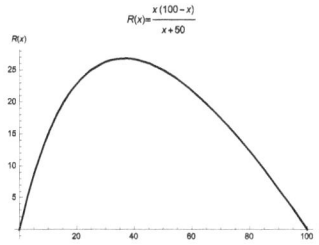

Figure 6.5: Plot of $R(x)$.

That is, $(md^2 - m\omega^2 + k - cd)\cos(\omega t + \phi) \equiv (c\omega - 2m\omega d)\sin(\omega t + \phi)$. But this identity can be true for all time t only if both sides vanish identically (since the sine and cosine functons do not coincide). This implies that the coefficients in front of the trigonometric functions must be zero, yielding

$$d = c/(2m), \text{ and } \omega = \sqrt{d^2 - \frac{cd - k}{m}} = \sqrt{4km - c^2}/(2m).$$

(b) See Fig. (6.6), with $A = 1, d = 1, \omega = \pi, \phi = 0$.

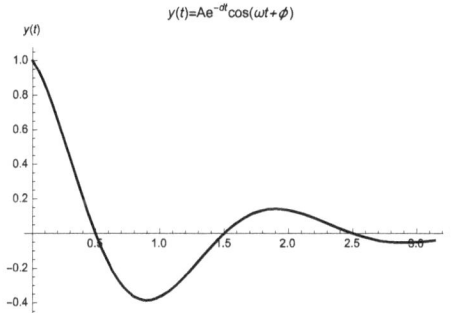

Figure 6.6: Plot of $y(t)$ with $A = 1, d = 1, \omega = \pi, \phi = 0$.

(c) When $c = 0$, the solution above reduces to $y(t) = A\cos(\omega t + \phi)$, where ω is now $\sqrt{k/m}$. This is indeed the solution of $m\dfrac{d^2 y}{dt^2} = -ky$, which is the equation for simple harmonic motion.

(d) When $k = 0$, the differential equation is $m\dfrac{d^2 y}{dt^2} = -c\dfrac{dy}{dt}$. If we let $\dfrac{dy}{dt} = z(t)$, then the equation becomes $m\dfrac{dz}{dt} = -cz(t)$, which has the well-known solution $z(t) = Ae^{-ct/m}$. That is, $\dfrac{dy}{dt} = Ae^{-ct/m}$, which has the solution $y(t) = D + Fe^{-ct/m}$, where D, F are constants determined by initial conditions.

Chapter 7

Integral Calculus

Recall: The average value of a function y over the range $a \leq x \leq b$ is given by

$$< y > \equiv \frac{\int_a^b f(x)dx}{b-a} . \tag{7.1}$$

---

1. The logistics equation is $dx/dt = Ax(B-x)$, with A and B positive constants. Re-arranging the terms, we have

$$\int dt = \int \frac{dx}{Ax(B-x)} = \int \frac{dx}{AB}\left(\frac{1}{x} + \frac{1}{B-x}\right)$$

which yields

$$ABt + k = \ln\frac{x}{|x-B|}$$

where k is an integration constant. At $t = 0$ we have $k = \ln\frac{x}{|x_0 - B|}$. Combining everything and simplifying, we get

$$x(t) = \frac{Bx_0 e^{ABt}}{x_0 e^{ABt} - |x_0 - B|}.$$

2. The temperature variation is described by the equation $f(t) = A + B\sin\omega t$, where A, B and ω have the values 27, 5, and 1.114, respectively.

   (a) The average temperature for $0 < t < 6$ is given by $T_{\text{ave}} = \frac{1}{6}\int_0^6 f(t)dt$.

   The integration of $f(t)$ with respect to t gives $T(t) = At - \frac{B}{\omega}\cos\omega t$. With the given values, we obtain $T_{\text{ave}} = 27.06°\text{C}$.

   (b) From part (a), we know that the equation for the average temperature is given by $T_{\text{ave}} = \frac{1}{T}\left[27t - \frac{5}{1.114}\cos 1.114t\right]\Big|_0^T$. For the duration of $0 < t < T$, the average temperature is found to be $29°\text{C}$. Therefore,

$$27T - \frac{5}{1.114}\cos 1.114T + \frac{5}{1.114} = 29T$$

$$\frac{5}{1.114}(1 - \cos 1.114T) = 2T$$

   The solution can be found by plotting $y = \frac{5}{2(1.114)}(1 - \cos 1.114T)$, and the line $y = T$. The intersection of the two lines will give us the solution, see Fig.(7.1). Since it is stated that $T > 1$, the solution is $T = 3.63$.

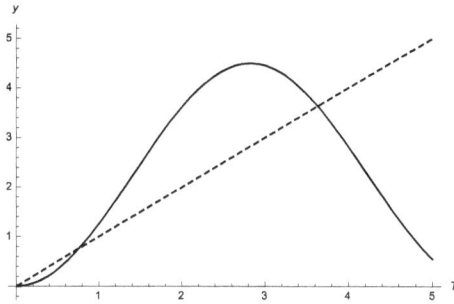

Figure 7.1: The solid line represents the plot for $y = \dfrac{5}{2(1.114)}(1 - \cos 1.114T)$ while the equation for the dashed line is $y = T$.

3. With $N(t) = N_0 \exp(-\lambda t)$, we have $dN(t)/dt = -\lambda N_0 \exp(-\lambda t)$. So the mean lifetime may be evaluated as follows

$$
\tau = \frac{\int_0^\infty t \left| \frac{dN(t)}{dt} \right| dt}{\int_0^\infty \left| \frac{dN(t)}{dt} \right| dt}
$$
$$
= \frac{\int_0^\infty t \exp(-\lambda t)\, dt}{\int_0^\infty \exp(-\lambda t)\, dt}
$$

Using integration by parts,

$$
\int_0^\infty t \exp(-\lambda t)\, dt = \left[\frac{-t}{\lambda} \exp(-\lambda t) \right]_0^\infty + \frac{1}{\lambda} \int_0^\infty \exp(-\lambda t)\, dt \ .
$$

The boundary term vanishes, so we easily obtain $\tau = 1/\lambda$.

4. The AC voltage at time t is $V(t) = V_0 \sin 2\pi f t$, where $f = 60$ Hz is the frequency and $V_0 = 240\sqrt{2}$ V. Each cycle corresponds to $0 < t < 1/2f$.

   (a) The mean voltage over one cycle, $< V >$ is 0, since the positive values and the negative values of the sine function will cancel each other over the course of one cycle.

   (b) From the definition of the mean in (7.1), the mean of the squared voltage over one cycle is given by (using the identity $2\sin^2\theta = 1 - \cos 2\theta$),

$$
\begin{aligned}
< V^2 > &= \frac{\int_0^{1/2f} V_0^2 \sin^2 2\pi ft\; dt}{1/2f} \\
&= \frac{1}{2}\frac{\int_0^{1/2f} V_0^2 (1 - \cos 4\pi ft)dt}{(1/2f)} \\
&= V_0^2 f\left[t - \frac{1}{4\pi f}\sin 4\pi ft\right]_0^{1/2f} \\
&= V_0^2/2 = (240)^2 V^2
\end{aligned}
$$

(c) The RMS voltage is then $\sqrt{< V^2 >} = 240$ V.

5. Using the given equation,

(a) Re-arranging, using partial fractions and integrating,

$$
\begin{aligned}
\int dt &= \int \frac{dv}{-v(1 + kv)} = -\int^V dv\left(\frac{1}{v} - \frac{1}{v + 1/k}\right) \\
t &= -\ln\frac{kV}{kV + 1} + C\,,
\end{aligned}
$$

where C is the integration constant. If the initial velocity at $t = t_0$ is v_0, then

$$
t - t_0 = \ln\frac{kv_0}{kv_0 + 1} - \ln\frac{kV}{kV + 1}
$$
$$
t = t_0 + \ln(v_0/V) + \ln[(kV + 1)/(kv_0 + 1)]
$$

(b) Re-arranging the terms above we obtain

$$
V = \frac{v_0 \exp(t_0 - t)}{1 + kv_0 - kv_0 \exp(t_0 - t)}
$$

(c) For $t \to \infty$, the term $\exp(t_0 - t) \to 0$. Therefore, $V \to 0$.

6. (a) Using the given wavefunction,

$$
\begin{aligned}
\int_0^L |\psi|^2 dx &= \int_0^L \frac{2}{L}\sin^2(\frac{n\pi x}{L})dx = \int_0^L \frac{1}{L}\left[1 - \cos(\frac{2n\pi x}{L})\right]dx \\
&= \frac{1}{L}\left[x - \frac{L}{2\pi n}\sin(\frac{2n\pi x}{L})\right]_0^L = 1\,.
\end{aligned}
$$

(b) Given that $<x> \equiv \int_0^L x|\psi|^2 dx$, for $n = 1$, we have

$$
\begin{aligned}
<x> &= \int_0^L \frac{2}{L} x \sin^2(\frac{\pi x}{L}) dx = \int_0^L \frac{1}{L} x \left[1 - \cos(\frac{2\pi x}{L})\right] dx \\
&= \frac{1}{L} \frac{x^2}{2}\Big|_0^L - \frac{1}{L} \int_0^L x \left[\cos(\frac{2\pi x}{L})\right] dx \\
&= \frac{L}{2} - \frac{1}{L} \int_0^L x \left[\cos(\frac{2\pi x}{L})\right] dx
\end{aligned}
$$

Using integration by parts,

$$
\int_0^L x \left[\cos(\frac{2\pi x}{L})\right] dx = \frac{L}{2\pi} x \sin\left(\frac{2\pi x}{L}\right)\Big|_0^L + \frac{L}{2\pi} \int_0^L \sin\left(\frac{2\pi x}{L}\right) dx = 0
$$

Hence, $<x> = L/2$.

(c) The probability of finding a particle in an infinitesimal interval dx is $P(x) = |\psi|^2 dx$. However, if we pinpoint an exact location, $dx = 0$. Therefore, the probability of finding a particle at one exact point will be zero.

7. For oxygen production,

$$
P(t) = \int \frac{dP(t)}{dt} dt = \int 2t dt = t^2 + A \ .
$$

Similarly, the consumption is

$$
C(t) = \int 0.5t^2 dt = \frac{0.5t^3}{3} + B
$$

where A and B are constants. Since $P(0) = C(0) = 0$, we obtain $P(t) = t^2$ and $C(t) = 0.5t^3/3$. To find the time taken for the two processes to balance each other, let $P(t) = C(t)$, which leads to $3t^2 - 0.5t^3 = 0$, that is, $t = 6$ (other than at $t = 0$).

8. The mass dm of an infinitesimal volume dV is given by $dm = \rho(r)dV$ where $\rho(r)$ is the density profile.

(a) In this case $\rho = \rho_0$ is a constant for for $r \leq R$. Hence,

$$
m = \int dm = \int_0^V \rho_0 dV = \rho_0 V = \frac{4}{3}\pi R^3 \rho_0 \ .
$$

(b) This time, $\rho = \rho_0 \frac{r}{R}$ for $r \leq R$. Since $V = 4\pi r^3/3$, so $\dfrac{dV}{dr} = 4\pi r^2$ and we can change the integration variable from V to r:

$$
\begin{aligned}
m &= \int_0^R \rho_0 \frac{r}{R} dV, \\
&= \int_0^R \frac{4\pi r^3}{R} \rho_0 dr \\
&= \pi \left[\frac{r^4}{R}\right]_0^R \rho_0 \\
&= \pi R^3 \rho_0 \ .
\end{aligned}
$$

9. $W = \displaystyle\int_{V_1}^{V_2} p\, dV$.

(a) For an ideal case, $p = nRT/V$. So at constant T, the work done by 1 mole of gas ($n = 1$) is

$$
W = \int_{V_1}^{V_2} \frac{RT}{V} dV = RT \ln\left(\frac{V_2}{V_1}\right).
$$

(b) For the Van der Waals case

$$
p = \frac{nRT}{V - nb} - \frac{an^2}{V^2} \ .
$$

The the work done by 1 mole of gas will then be

$$
\begin{aligned}
W &= \int_{V_1}^{V_2} \left[\frac{RT}{V - b} - \frac{a}{V^2}\right] dV \\
&= \left[RT \ln(V - b) + \frac{a}{V}\right]_{V_1}^{V_2} \\
&= RT \ln\left(\frac{V_2 - b}{V_1 - b}\right) + \frac{a}{V_2} - \frac{a}{V_1} \ .
\end{aligned}
$$

10. The center of mass of a rod is given by

$$
r_{cm} = \frac{\int r\rho\, dr}{\int \rho\, dr} \tag{7.2}
$$

where $\rho(r)$ is the line density of the rod.

(a) In this case $\rho(r) = \rho_0$ for $0 \leq r \leq L$ and zero otherwise. So the center of mass is at

$$
r_{cm} = \frac{\int_0^L r\rho_0\, dr}{\int_0^L \rho_0\, dr} = \frac{(r^2\rho_0)/2|_0^L}{r\rho_0|_0^L} = L/2
$$

(b) The non-vanishing line density is now

$$\rho(r) = \frac{r\rho_0}{L}, 0 \le r \le L \qquad (7.3)$$

Hence the center of mass is

$$r_{cm} = \frac{\int_0^L (r^2 \rho_0/L) dr}{\int_0^L r\rho_0/L dr} = \frac{(r^3 \rho_0)/3|_0^L}{r^2 \rho_0/2|_0^L} = 2L/3$$

11. Given $\ddot{y} = -g$ with $v(t) = \dot{y}(0) = u > 0$ and $y(0) = 0$.

   (a)

$$v(t) = \dot{y} = \int_0^t \ddot{y} dt = -gt + C$$

   where C is the integration constant. Since $v(0) = u$, we get $v(t) = u - gt$.

   (b) Integrating again,

$$y(t) = \int \dot{y} dt = \int_0^t (u - gt) dt = ut - \frac{1}{2}gt^2 + B,$$

   with B being a constant. Given that $y(0) = 0$, we obtain $y(t) = ut - gt^2/2$.

   (c) At the maximum height the velocity is momentarily zero. So setting $dy/dt = \dot{y} = 0$ gives us $0 = u - gt$. Therefore, the maximum height corresponds to a time $t = u/g$.

   (d) From part (c) we know that the maximum height is reached when $t = u/g$. Substituting this into the expression for the object height $y(t) = ut - gt^2/2$, gives $h_{max} = u^2/2g$.

   (e) From $y(t) = t(u - gt/2)$, the time taken for the particle to return to $y = 0$ is determined to be $2u/g$.

   (f) The particle returns to $y = 0$ at time $t = 2u/g$. Using the expression obtained for the velocity of the particle, $v(t = 2u/g) = -u$.

12. Water is pumped in at a constant rate, $\dfrac{dV}{dt} = 3$ cm^3 s^{-1}, where t is the time and V the volume.

   (a) We get $V = \int dV = \int 3dt = 3t + C$ where C is the integration constant. Since $V = 200$ at $t = 10$, so $C = 170$ and $V = 3t + 170$. Hence $V(t = 1) = 173$, and using $V = 4\pi r^3/3$ gives us $r = 3.457$ cm.

   (b) From the expression $V = 4\pi r^3/3$, we get $dV/dr = 4\pi r^2$. We also have from the chain rule,

$$\frac{dV}{dt} = \left(\frac{dV}{dr}\right)\left(\frac{dr}{dt}\right) = 4\pi r^2 \left(\frac{dr}{dt}\right)$$

Therefore,

$$\frac{dr}{dt} = \frac{3}{4\pi r^2} = 0.02 \text{ cm s}^{-1}$$

where we used $r = 3.457$ from the first part.

(c) The surface area is $A = 4\pi r^2$. So

$$\frac{dA}{dt} = \frac{dA}{dr} \times \frac{dr}{dt} = (8\pi r)\frac{dr}{dt} \ .$$

Using the previously obtained values for r and dr/dt, the rate of change of the surface area is $1.737 \text{ cm}^2 \text{ s}^{-1}$.

13. The net inflow is $dV/dt = 10t - t^2$.

(a) The volume V of cement in the cone at a given time t can be found by integration,

$$V = \int \frac{dV}{dt} dt = 5t^2 - t^3/3 + 10$$

where we fixed the constant of integration using $V(0) = 10$.

(b) A plot of $V(t)$ against t is shown Figure 7.2. The local minimum is found at $t = 0$, with $V = 10$, while the maximum occurs at $t = 10$ with $V = 530/3$.

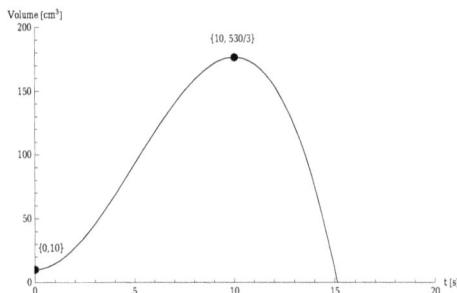

Figure 7.2: The graph of $V(t)$ against t. The dots indicate the local extrema.

(c) The local maximum value V_{max} can also be found by setting $dV/dt = 10t - t^2 = 0$, and ensuring that $d^2V/dt^2 < 0$. From the first order derivative, we obtain the solutions $t = 0$ and $t = 10$. From $d^2V/dt^2 = 10 - 2t$, we can see that at $t = 10$, the value of $d^2V/dt^2 < 0$, which means that the local maximum value is at $t = 10$, with $V_{\text{max}} = 530/3$. From the figure above we see that the local maximum is a global maximum.

(d) The local minimum value V_{min} should satisfy the conditions $dV/dt = 0$ and $d^2V/dt^2 > 0$. At $t = 0$, the the conditions are fulfilled.

(e) The local minimum value is found to be $V = 10$ at $t = 0$. The time at which the volume goes below the local minimum can be determined by setting $V = 5t^2 - \dfrac{t^3}{3} + 10 < 10$, resulting in $t^2(5 - \dfrac{t}{3}) < 0$, that is $t > 15$ s.

(f) Since the flow rate of the cement (with time) out of the cone is quadratic while the input flow is linear, the outflow will eventually exceed the inflow. Therefore, there will be a time when there is no cement in the cone, as seen also from the figure.

(g) With $V = h^3/20$ cm^3, we have

$$\frac{dV}{dt} = \left(\frac{dV}{dh}\right)\left(\frac{dh}{dt}\right)$$
$$10t - t^2 = \frac{3h^2}{20}\left(\frac{dh}{dt}\right)$$
$$\Rightarrow \frac{dh}{dt} = \frac{20(10t - t^2)}{3h^2} \ .$$

At $t = 5$, the volume of cement V is $280/3$ cm^3, corresponding to $h = 12.313$ cm. Substituting these values of t and h into the expression for dh/dt, we obtain $dh/dt = 1.1$ cm s^{-1}.

Did You Know?

The integration symbol $\displaystyle\int$ arises from an elongated letter S.

The mathematician Leibniz, who invented calculus independently of Newton, chose that symbol to represent the fact that integration may be viewed as an infinite **S**um, such as when one calculates the area under a curve by a sum of thin rectangles whose width is then taken to zero.

Free E-books

The E-book editions of *Handbook of Mathemtics* and other books are available for download from
www.simplicitysg.net/books/free-ebooks

Chapter 8

Vectors and Complex Numbers

1. Speed is the magnitude of the velocity vector.

   (a) For Al, the speed is $|\vec{v}_a| = \sqrt{3^2 + 4^2} = 5$ km/h. Similarly, for Bel, the speed is $|\vec{v}_b| = \sqrt{3^2 + (-4)^2} = 5$ km/h, too.

   (b) The relative velocity of Al to Bel is $\vec{v}_a - \vec{v}_b = (3-3)\vec{i} + (4+4)\vec{j} = \left(0\vec{i} + 8\vec{j}\right)$ km/h.

   (c) The angle between the two vectors can be determined through their scalar product $\vec{v}_a \cdot \vec{v}_b = |\vec{v}_a||\vec{v}_b| \cos\theta$. For the LHS, we have $\vec{v}_a \cdot \vec{v}_b = 9 - 16 = -7$. For the RHS, we have $25\cos\theta$. So $\theta = \arccos(-7/25) = 106.26°$ or $257.74°$. Of course, once we plot out the two vectors, it is quite clear, by inspection, that the angle is the smaller one, that is, $\theta = 106.26°$.

   Note: More generally, to decide which angle is the correct one without resorting to plots, we can compute their cross product: $\vec{v}_a \times \vec{v}_b = |\vec{v}_a||\vec{v}_b| \sin\theta\, \vec{n}$, where $\vec{n}$ is the unit vector perpendicular to the plane defined by $\vec{v}_a$ and $\vec{v}_b$ with its direction given by the right-hand rule (see page 126 in the book). In our case, $\vec{v}_a \times \vec{v}_b = -24\vec{k}$. By the right-hand rule, $\vec{n}$ is also $-\vec{k}$. Hence, $24 = 25\sin\theta$, telling us that $\sin\theta$ is positive, which means that we must choose $\theta = 106.26°$.

2. (a) Two vectors can always be put in a common plane. Then, they define a parallelogram in this plane, which can be viewed as two triangles, each of area $(ab\sin C)/2$. The area of the parallelogram is thus given by the magnitude of the cross product $|\vec{a} \times \vec{b}| = |16\vec{i} + 7\vec{j} - 10\vec{k}| = \sqrt{16^2 + 7^2 + 10^2} = \sqrt{405} = 9\sqrt{5} \approx 20.12$.

   (b) Given three vectors $\vec{a}, \vec{b}, \vec{c}$, they define a parallelepiped with volume ('base area times vertical height'), $V = |(\vec{a} \times \vec{b}) \cdot \vec{c}|$. For our case, we have $V = |(16\vec{i} + 7\vec{j} - 10\vec{k}) \cdot (2\vec{i} + \vec{j} - \vec{k})| = 32 + 7 + 10 = 49$.

3. (a) Identifying North direction with the unit vector $\vec{j}$, East direction with the unit vector $\vec{i}$, and Al's house as the origin of the coordinates, the position vector of the school is $\vec{r}_1 = 30\vec{i} + 8\vec{j}$, and the position vector of Neo's house is $\vec{r}_2 = 10\vec{i} + 15\vec{j}$. The relative position vector between Neo's house and the school is then $\vec{r}_3 = \vec{r}_1 - \vec{r}_2 = 20\vec{i} - 7\vec{j}$, that is, the school is 20 km to the East, and 7 km to the South of Neo's house. See also Fig. (8.1) for illustration.

   (b) The shortest distance is a straight line, which is equal to the magnitude of $\vec{r}_3$. That is, $|\vec{r}_3| = \sqrt{20^2 + 7^2} = \sqrt{449} = 21.19$ km.

(c) We want to compute the distance between Neo's house and where he is picked up. Since he is picked up at the point on the road where it is shortest distance from his house, it is a straight line connecting his house to that point, and from trigonometry, we know that it is perpendicular to the road. The distance is then $x = |\vec{r}_2| \sin\theta$, where θ is the angle between $\vec{r}_1$ and $\vec{r}_2$. Its numerical value is $x = |\vec{r}_2| \sin\theta = \dfrac{|\vec{r}_2||\vec{r}_1|\sin\theta}{|\vec{r}_1|} =$

$\dfrac{|\vec{r}_2 \times \vec{r}_1|}{|\vec{r}_1|} = \dfrac{370}{\sqrt{964}} = 11.92\,\text{km}$. See Fig. (8.1) for illustration.

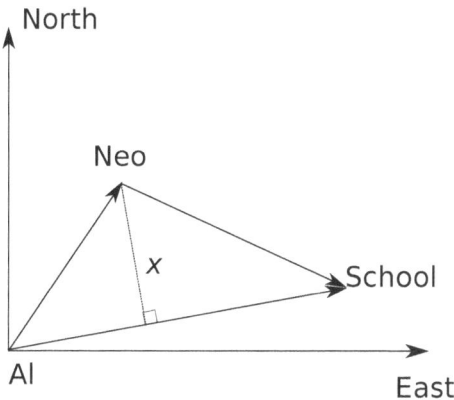

Figure 8.1: Figure (drawn not to scale) depicting the relevant positions and distances in Ex. (8.3).

4. (a) The total velocity vector of Romeo is obtained by adding the drift of the river to the thrust generated by his own rowing, $\vec{v} = 0.25\vec{i} + 1\vec{j}$. As in the previous problem, we identify North with the unit vector $\vec{j}$ and East with the unit vector $\vec{i}$. Then, his resultant speed is $|\vec{v}| = \sqrt{0.25^2 + 1^2} = \sqrt{17}/4\,\text{m/s}$. Meanwhile, it takes him $50/1 = 50$ s to reach the North bank, and during this 50 s, the river has drifted him $0.25 \times 50\,\text{s} = 12.5\,\text{m}$ towards East. The angle as measured from his original position is then $\theta = \arctan(12.5/50) = 14.04°$.

   (b) By intuition, he has to row towards some direction West of North such that the river will bring him exactly to Juliet when he reaches the North bank, see Fig.(8.2) for the velocity vectors. The angle is therefore $\theta = \arcsin(0.25/1) = 14.48°$.

   (c) In the above situation, Romeo effectively travels 50 m along the straight line connecting his original position to Juliet. His speed is $0.25/\tan\theta$ and thus the time taken is $50/(0.25\tan\theta) = 200\tan\theta = 51.64\,\text{s}$.

5. (a) The cart moves along the horizontal road $(\vec{i})$, and the force $\vec{F}$ of $5\,\text{N}$

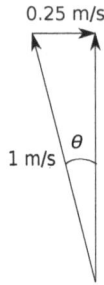

Figure 8.2: Velocity vector triangle for Q4(b).

magnitude is applied at an angle of $30°$ to the road. Hence the component in the direction of motion is $\vec{F} \cdot \vec{i} = 5\cos(30°) = 5\sqrt{3}/2\,\text{N}$.

(b) The force along the normal component $(\vec{j})$ is $\vec{F} \cdot \vec{j} = 5\sin(30°) = 5/2\,\text{N}$. The total force is then $\vec{F} = 5/2\vec{j} + 5\sqrt{3}/2\vec{i}\,\text{N}$.

(c) $W = \vec{F} \cdot \vec{x} = 5\sqrt{3}/2 \times 10 = 25\sqrt{3}\,\text{N}$.

6. With $\psi(x) = e^{-ikx}$, we have $\dfrac{d^2}{dx^2}\psi(x) = (-ik)^2\psi(x)$. Hence, if we have
$$E = \frac{\hbar^2 k^2}{2m}, \quad -\frac{\hbar^2}{2m}\frac{d^2}{dx^2}\psi(x) = E\psi(x).$$

7. With $\psi(x) = e^{-ikx}$, we have $\dfrac{d}{dx}\psi(x) = (-ik)\psi(x)$. Hence, if we have $\hbar k = p$,
$$i\hbar\frac{d}{dx}\psi(x) = p\psi(x).$$

8. (a) $e^{iA} \times e^{iB} = e^{i(A+B)}$. By Euler's identity, the LHS is $(\cos A + i\sin A) \times (\cos B + i\sin B) = \left[\cos A \cos B - \sin A \sin B\right] + i\left[\sin A \cos B + \cos A \sin B\right]$, while the RHS is $\cos(A + B) + i\sin(A + B)$. Equating the real and imaginary parts, we have $\cos(A + B) = \cos A \cos B - \sin A \sin B$, and $\sin(A + B) = \sin A \cos B + \cos A \sin B$, the sum rules for sine and cosine.

(b) We have $e^{i4x} = \cos(4x) + i\sin(4x)$, and also $e^{i4x} = \left(e^{ix}\right)^4 = (\cos x + i\sin x)^4$. Expanding the last expression and comparing the the real and imaginary parts with the first expression we obtain $\cos(4x) = \cos^4 x + \sin^4 x - 6\sin^2 x \cos^2 x$ and $\sin(4x) = 4\sin x \cos^3 x - 4\sin^3 x \cos x$.

(c) By using $\cos(0) = \sin(\pi/2) = 1$, $\cos(\pi/4) = \sin(\pi/4) = 1/\sqrt{2}$, $\cos(\pi/2) = \sin(0) = 0$, and $\cos(\pi) = \sin(3\pi/2) = -1$, and the fact that cosine and sine are periodic functions, that is, $\cos(x + 2\pi n) = \cos(x)$, $\sin(x + 2\pi n) = \sin(x)$ for any integer n, one can easily verify the results in the last part.

www.ingramcontent.com/pod-product-compliance
Lightning Source LLC
Chambersburg PA
CBHW071123210326
41519CB00020B/6404